In Vitro Culture of Higher Plants

by

R.L.M. PIERIK

Department of Horticulture, Agricultural University
Wageningen, The Netherlands

1987 **MARTINUS NIJHOFF PUBLISHERS**
a member of the KLUWER ACADEMIC PUBLISHERS GROUP
DORDRECHT / BOSTON / LANCASTER

UNIVERSITY LIBRARY

20 JUL 1988

LANC

Distributors

for the United States and Canada: Kluwer Academic Publishers, P.O. Box 358, Accord Station, Hingham, MA 02018-0358, USA
for the UK and Ireland: Kluwer Academic Publishers, MTP Press Limited, Falcon House, Queen Square, Lancaster LA1 1RN, UK
for all other countries: Kluwer Academic Publishers Group, Distribution Center, P.O. Box 322, 3300 AH Dordrecht, The Netherlands

Library of Congress Cataloging in Publication Data

```
Pierik, R. L. M.
   In vitro culture of higher plants.

   Rev. translation of: Plantenteelt in kweekbuizen.
   Bibliography: p.
   Includes index.
   1. Plant propagation--In vitro.  2. Plant tissue
culture.  3. Plant cell culture.  I. Title.
SB123.6.P4813  1987        582'.00724        87-11235
```

ISBN 90-247-3531-9 (paperback)
ISBN 90-247-3530-0 (hardback)

Copyright

© 1987 by Martinus Nijhoff Publishers, Dordrecht.

All rights reserved. No part of this publication may be reproduced, stored in a retrieval system, or transmitted in any form or by any means, mechanical, photocopying, recording, or otherwise, without the prior written permission of the publishers,
Martinus Nijhoff Publishers, P.O. Box 163, 3300 AD Dordrecht,
The Netherlands.

PRINTED IN THE NETHERLANDS

87   08274

Contents

IV

1. Preface

When the first Dutch edition of this book appeared in 1975 it was not anticipated that it would arouse so much interest. The rapid development of this subject pursuaded the author to produce an extensively revised second edition in 1985. This third edition has been further revised and extended by 20% to accomodate the recent advances, including those presented at the 6th Plant Tissue and Cell Culture Congress in Minnesota (1985) and the 12th Horticultural Congress in California (1986).

Since 1975 in vitro culture of higher plants has shown a spectacular development. In scientific laboratories methods have been developed for the culture of plants, seeds, embryos, shoot tips, meristems, tissues, cells and protoplasts on sterile nutrient media, resulting in the production and regeneration of viable individuals of many plant species. Since 1980 there has been an explosion in the development of genetic manipulation and biotechnology techniques.

The production of higher plants in vitro, one of the most important aims of tissue culture, strongly appeals to the imagination. But to philosophize about the interesting aspects of producing a plant from a cell, tissue, etc. does not take us very far. The practical person, such as an agriculturist is more interested in knowing whether vegetative propagation can be satisfactorily achieved in this way, especially in cases where cloning by current methods such as by cuttings, division, budding and grafting are difficult or even impossible.

This book has been written in response to the practical problems and questions encountered by the author working in a horticultural laboratory. There are two aspects to these practical problems. Firstly, it was necessary to write a book suitable for university students, as an introduction to the in vitro culture of higher plants. Secondly, tissue culture laboratories (research and applied) required a practical handbook especially for technicians and research assistants. As a result of consultation with numerous colleagues in teaching and research as well as those involved in

1

practical agriculture and horticulture, the author has attempted to answer these demands.

The author is indebted to Prof. Dr. Ir. J. Doorenbos, and Dr. C.J. Gorter for their critical appraisal of the first edition of this book.

To avoid errors and omissions during preparation of the second edition I am grateful to colleagues who accepted my invitation, for them to critically read specific Chapters: Dr. Ir. J.B.M. Custers (Chapter 16), Drs. F. Quak (Chapter 19), Ir. J.A.J. van der Meys and Dr. G. Staritzky (Chapters 20 and 21), Prof. Dr. Ir. J.G.Th. Hermsen (Chapters 22, 23, and 24), Dr. J.J.M. Dons (Chapter 24), Dr. Ir. M. Koornneef (Chapter 24) and Dr. Ir. C. Broertjes, Prof. Dr. Ir. J. Doorenbos (Chapter 25). Many others sent me criticisms and additions for which I am most grateful. The author hopes that in subsequent years, the readers of this third edition will supply further critical comments, suggestions, omissions or errors.

I am especially grateful to the technicians and research assistants who have worked with me, over the years, in the Department of Horticulture of the Agricultural University, Wageningen, without whose contributions the research presented in this book would not have been possible. Numerous M. Sc. students have also made a great contribution to this new edition through their research during the last years of their study. This research has been a source of inspiration and the foundation stone for writing this third edition.

My present technical assistants H.H.M. Steegmans and P.A. Sprenkels have given much help over many years. Their critical observations and additions have been of great value during the preparation of this book. Thanks go to my wife for much of the typing and M.A. Ruibing for corrections during the final stages of preparation. Without the support of my sons A.J. Pierik and L.M. Pierik in the use of the computer/word-processor the appearance of this book would have certainly been delayed several months.

I am indebted to C.I. Kendrick for the English translation, and greatly appreciate Dr. P. Debergh critically appraising the manuscript in its final phase of production.

Wageningen the Netherlands, April 1987   Prof. Dr. Ir. R.L.M. Pierik

2. Introduction

2.1. Outline

When people talk about plant culture, they usually mean the growing of plants in pots, frames, greenhouses or in the field. Plant culture is further subdivided into the disciplines of agriculture, horticulture, tropical agriculture, forestry, and plant breeding. In 1904 Hännig developed a new method of plant culture called embryo culture. He isolated (immature) embryos in vitro and obtained viable plantlets of several members of the *Cruciferae*. Many types of culture have become popular since 1920, such as: in vitro sowing of orchid seeds, callus culture, organ culture etc. (see Chapter 3). After 1945 all these diverse types of culture were grouped together under the collective term plant tissue culture, an unlikely name for such a diverse group of cultures.

To avoid confusion the title of this book has been chosen as: In vitro culture of higher plants. This is defined as the culture on nutrient media under sterile conditions, of plants, seeds, embryos, organs, explants, tissues, cells and protoplasts of higher plants. These techniques are characterised by:

1. They occur on a micro-scale i.e. on a relatively small surface area.
2. The environmental conditions are optimized with regards to physical as well as nutritional and hormonal factors.
3. All micro-organisms (fungi, bacteria and viruses), as well as other pests of higher plants (insects, nematodes) are excluded.
4. The normal pattern of plant development often breaks down, and an isolated tissue can give rise to a callus or can develop in many unusual ways (e.g. organ formation, somatic embryogenesis).
5. The ability to grow protoplasts or individual cells enables manipulations which were previously impossible.

6. The name in vitro culture (in vitro literally meaning 'in glass') came into being because, initially at least, glass vessels were used for the culture.

Since this book actually contains descriptions of special types of in vitro culture (Fig. 2.1.), it can be described as a practical manual or methodology for in vitro culture techniques. Typical methodological aspects of in vitro culture are given in Chapters 5 to 14: laboratory set up, preparation of nutrient media, closure of test tubes and flasks, growth and sterilization of the plant material, isolation, inoculation and subculturing, mechanization, the influence of the starting material and physical factors on the growth and development etc.

In vitro culture also has a definite role to play, and is a tool for achieving things impossible in vivo. Returning to practical considerations: why is in vitro culture practised? This question is extensively covered in Chapters 16 to 25, both with regards to practical applications in agriculture, horticulture and forestry as well as more fundamental research.

A list of different types of culture and their possible application is given below with the relevant Chapter number in brackets. Technical terms used in the table are defined in Section 2.2. Details in relation to each culture and its practical applications are covered in the related Chapters.

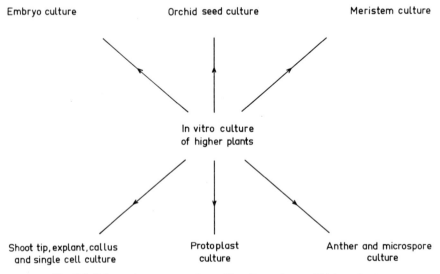

Fig. 2.1. Schematic representation of in vitro culture of higher plants.

Type of culture	Purpose
Embryo culture (16)	Shortening of the breeding cycle Prevention of embryo abortion Overcoming incompatibility Production of haploids As a source of callus formation
Orchid seed culture (17)	Shortening of the breeding cycle Replacing symbiosis (mycorrhiza) Excluding competition with other micro-organisms
Meristem culture (18, 19)	Elimination of pathogens (viruses, fungi, bacteria) Vegetative propagation of orchids through proto- corms Cloning of plants other than orchids Storage of disease-free plants Phytosanitary transport Germplasm collection
Shoot tip and single-node culture (18, 20)	Orchid propagation Axillary branching as a tool to clone plants Cryo-preservation to create gene banks
Explant (without pre-existing buds) culture (19, 20, 25)	Adventitious organ formation for cloning plants Obtaining disease-free plants Production of solid mutants (mutation breeding) Isolation of mutants Solving the chimera problem Obtaining polyploids
Callus, suspension and single cell culture (20, 21, 25)	Cloning of plants through organ and embryo forma- tion Creation of genetic variants Obtaining virus-free plants As a source for protoplast production Starting material for cryo-preservation Production of secondary metabolites Biotransformation
Anther and microspore culture (23)	Production of haploids and to obtain homozygotes As a starting point for mutation induction Creation of all-male plants As a tool in genetic manipulation To breed at lower ploidy levels
Culture of ovules and excised flowers (16, 22)	Overcoming incompatibility To prevent precocious abscission of flowers Achievement of test-tube fertilization
Protoplast culture (24)	Somatic hybridization Creation of cybrids Transplantation of nuclei, (fragments of) chromos- omes and organelles Transformation studies Creation of genetic variants

Culture of protoplasts, cells, tissues and organs (25)	As a tool in phytopathological research
	– Virus penetration and replication
	– Culture of obligate parasites
	– Host-parasite interactions
	– Culture of nematodes (excised root cultures)
	– Testing of phytotoxins
	– Nodulation studies
	As a tool in plant physiological research
	– Cell cycle studies
	– Metabolism
	– Nutritional studies
	– Morphogenetical and developmental studies

Vegetative propagation in vitro (also called micropropagation) will be dealt with in this book in greater detail since it has yielded results, which are of enormous importance for agriculture, horticulture and forestry. It is especially notable in horticulture, where people quickly respond to the results of research on in vitro culture. It is also particularly striking that horticulturists very frequently use in vitro culture as a means of vegetative propagation. Due to in vitro culture considerable saving in space (greenhouses) and, therefore, in energy, have drastically changed the production of vegetatively propagated plants. This is illustrated in Section 3.2. and Chapter 26.

As well as their practical applications, plant cell, tissue and organ culture have also made an important contribution to our fundamental knowledge of the cell. The cell theory of Schwann and Schleiden (1838–39), in which the cell was described as the smallest biological unit that could be considered totipotent has been substantiated by tissue culture: a single cell is capable of developing into a complete plant.

After the successful isolation of viable protoplasts from which whole plants could be regenerated, an alternative (parasexual hybridization or somatic hybridization) to conventional propagation techniques (sexual hybridization or crossing) was created. The possibility of forming individuals after protoplast fusion means that it is in principle possible to overcome natural genetic barriers. Somatic hybridization makes the creation of cytoplasmic hybrids (cybrids) possible. Techniques such as transplantation of nuclei or (parts of) chromosomes and organelles have been investigated by geneticists, plant breeders and molecular biologists. At this moment cell culture is of great importance in biotechnology: researchers are striving to grow plant cells like micro-organisms to obtain industrial production of secondary metabolites.

6

2.2. Abbreviations and glossary

ABA	abscisic acid
BA	6-benzylaminopurine
°C	degrees Celsius
CCC	(2-chloroethyl)trimethylammonium chloride
2,4-D	2,4-dichlorophenoxyacetic acid
DMSO	dimethylsulphoxide
DNA	deoxyribonucleic acid
EDTA	ethylenediaminetetraacetate
ELISA	enzyme-linked immunosorbent assay
GA	gibberellic acid
IAA	indole-3-acetic acid
IBA	indole-3-butyric acid
2iP	6-(γ,γ-dimethylallylamino)purine
MS	Murashige and Skoog (1962)
NAA	α-naphthaleneacetic acid
PBA	6-(benzylamino)-9-(2-tetrahydropyranyl)-9H-purine
PEG	polyethylene glycol
PVP	polyvinylpyrrolidone
RNA	ribonucleic acid
TIBA	2,3,5-triiodobenzoic acid
UV	ultraviolet (light)
v/v	volume/volume (concentration)
W	Watt
w/v	weight/volume (concentration)
Zeatin	6-(4-hydroxy-3-methyl-2-butenylamino)purine

Abscisic acid: Abbreviated by ABA, plant hormone that plays a role in dormancy and senescence.

Adventitious: Development of organs (roots, buds, shoots, flowers, etc.) or embryos (embryo-like structures) from unusual points of origin, including callus (Fig. 2.2.). If organs develop from organ initials, organ primordia, or embryos develop from zygotes, the term adventitious cannot be used.

Agar: A vegetable product (made from algae) used to solidify nutrient media.

Alcohol: Ethyl alcohol (C_2H_5OH), also called ethanol.

7

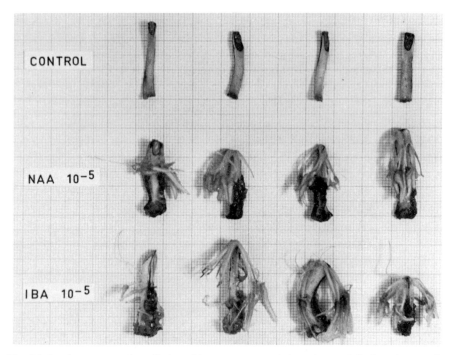

Fig. 2.2. In vitro regeneration of adventitious roots on petiole explants of *Gerbera jamesonii* in vitro. Control represents the absence of auxin. A concentration of 10^{-5} corresponds to $10\,mg\,l^{-1}$. Rooting only occurs when auxin is applied. Photograph taken 8 weeks after isotation.

Amino acids: Group of organic compounds, which among other things are the constituents of the biosynthesis of proteins.

Androgenesis: Male parthenogenesis. The development of a haploid individual from a pollen grain.

Aneuploid: A cell in which the number of chromosomes deviates from x (the haploid number) or multiples of x.

Anti-oxidants: A group of chemicals which prevents oxydation, e.g. vitamin C, citric acid.

Antisepsis: Process or principles using antiseptics.

Antiseptic: Counteracting sepsis, especially by preventing growth of micro-organisms.

8

Apical dominance: The phenomenon where the terminal bud of a shoot suppresses the outgrowth of the axillary buds.

Apical meristem: A group of meristematic cells at the apex of a root or a shoot which by cell division produces the precursors of the primary tissues of root or shoot.

Apomixis: Substitution of sexual reproduction by different types of asexual reproduction, where no gamete fusion occurs.

Asepsis: Absence of micro-organisms.

Aseptic: Free from all micro-organisms (fungi, bacteria, yeasts, viruses, mycoplasmas, etc.), sterile.

Asymbiotic: Not in symbiosis with micro-organisms. See also under symbiosis.

Autoclave: Apparatus in which media, glassware, etc. are sterilized by steam under pressure.

Autotrophic: Not requiring organic substances.

Auxins: Group of plant hormones (natural or synthetic), which induce cell elongation, or in some cases cell division; often inducing adventitious roots and inhibiting adventitious buds (shoots).

Axenic: Isolated from other organisms.

Axillary: Originating in the axils of the leaves.

Binocular: Stereomicroscope magnifying by 5–20 times.

Biosynthesis: Synthesis of compounds by the plant and cells.

Callus: Actively dividing non-organized tissues of undifferentiated and differentiated cells often developing from injury (wounding) or in tissue culture.

Cambium: Dividing tissue which in stems forms wood and bark.

Carbohydrates: A group of compounds serving as a source of energy (e.g. glucose, sucrose, fructose, etc.).

Casein hydrolysate: Mixture of compounds (in particular amino acids), manufactured from casein.

Cell culture: The growing of cells in vitro.

Cell generation time: The interval between consecutive cell divisions.

Cell line: Cells (originating from a primary culture) successfully subcultured for the first (second, etc.) time.

Cell strain: As soon as selection or cloning of cells with specific properties or markers starts from a primary culture or a cell line, we speak of a cell strain.

Chimera: A plant which contains groups (layers) of cells which are genetically dissimilar.

Clone: A group of cells, tissues, or plants which are in principle genetically identical. A clone is not necessarily homogeneous.

Coconut milk: Liquid endosperm of the coconut.

Contaminant: Micro-organism.

Culture room: Room for maintaining cultures with controlled light, temperature and humidity (Fig. 2.3.).

Cybrid: A cytoplasmic hybrid, originating from the fusion of a cytoplast (the cytoplasm without nucleus) with a whole cell.

Cytokinins: A group of plant hormones (natural or synthetic) which induce cell division and often adventitious buds (shoots) and in most cases inhibit adventitious root formation; cytokinins decrease apical dominance.

Dedifferentiation of cells: Reversion of differentiated to non-differentiated cells (meristematic).

Deionized water: Water which is free of inorganic compounds.

Detergent: Substance which lowers the surface tension of a solution; added to improve contact between plant and sterilizing agent.

Fig. 2.3. Culture room in a commercial tissue culture laboratory in California, U.S.A.

Development: The passage of a cycle from seed to seed or from organ initial to senescence; also changes in form of a plant by growth and differentiation.

Differentiation: The development of cells or tissues with a specific function and/or the regeneration of organs or organ-like structures (roots, shoots, etc.) or (pro)embryos.

Differentiation of cells: Cells taking on (a) specific function(s).

Dihaploid: This is an individual (denoted by $2n = 2x$) which arises from a tetraploid ($2n = 4x$) (Hermsen, 1977).

Diplophase: Phase with 2n chromosomes.

Diploid: A nucleus is diploid if it contains twice the base number (x) of chromosomes (Hermsen, 1977). The genome formula is $2n = 2x$.

Disease-free: This should be interpreted to mean 'free from any known diseases' as 'new' diseases may yet be discovered present.

Disease-indexing: Disease-indexed plants have been assayed for the presence of known diseases according to standard testing procedures.

Disinfection: Killing of micro-organisms.

Distilled water: Water produced by distillation containing no organic or inorganic compounds.

Doubling time: Term used in tissue culture and shoot propagation for the time necessary to double the number of cells/shoots in vitro.

EDTA: Ethylenediamine tetraacetic acid, a chelating compound, in which iron is so bound that it is still available for the plant.

Embryo abortion: Death of an embryo.

Embryogenesis: Process by which an embryo develops from a fertilized egg cell or asexually from a (group of) cell(s).

Embryoid: Plantlet, embryo-like in structure, produced by somatic cells in vitro; also adventitious embryo developing in vitro by vegetative means.

Embryo culture: The culture of embryos on nutrient media.

Endomitosis: Doubling of the number of chromosomes without division of the nucleus, resulting in polyploidy.

Endopolyploidy: The production of polyploid cells as a result of endomitosis.

Epigenetic variation: Non-hereditory variation which is at the same time reversible; often the result of a changed gene expression.

Erlenmeyer flask: Flat bottomed, conical shaped flask.

Excise: Cutting out (with knife, scalpel, etc.) and preparing a tissue, organ, etc. for culture.

Explant: An excised piece of tissue or organ taken from the plant, used to initiate a culture.

Filter sterilization: Process of sterilizing a liquid by passage through a filter, with pores so small that they are impervious to micro-organisms.

Flaming: A technique for sterilizing instruments by heating in a flame after dipping in alcohol.

Gibberellins: Group of plant hormones which induce, among other things, cell elongation and cell division.

Growth room: See culture room.

Habituation: The phenomena that, after a number of subcultures, cells can grow e.g. without the addition of hormones, although this was originally necessary.

Haploid induction: The stimulation of normal male and female gametes (after meiosis) to grow autonomically (Hermsen, 1977).

Haploid plant: Plant with half the number of chromosome (denoted by n) due to reduction division of the diploid ($=2n$) (de Fossard, 1976).

Hardening off: Gradual acclimatization of in vitro grown plants to in vivo conditions e.g. gradual decrease in humidity.

Heterokaryon: Cell with two or more different nuclei as a result of cell fusion.

Heterotrophic: Requiring organic substances.

Homokaryon: Cell with two or more identical nuclei as a result of fusion.

Hormone: Organic substance which is produced within a plant and which will at low concentrations promote, inhibit, or quantitatively modify growth, usually at a site other than its place of origin.

Hybridization of cells: The fusion of two or more dissimilar cells, resulting in the formation of a hybrid cell.

Induction: The initiation of a particular process, resulting in the development of organs (e.g. roots, shoots, flowers).

Infection: Communication of disease by micro-organisms.

Initial: Group of cells which serve as the precursor of an organ (leaf, root, bud).

Initiation: The formation of a structure or an organ e.g. a root or a shoot primordium.

Inoculate: Place in or on a nutrient medium (Fig. 2.4.).

Inoculation cabinet: Small room or cabinet for incoulation, often with a current of sterile air.

Intercalary: Between two adjacent buds on a stem, within an internode.

Intergeneric: Used in the case of a cross between two different genera.

Interspecific: Used in the case of a cross between two different species.

Juvenile phase: The period in the life of a plant during which no flowering can be induced. During the juvenile phase a plant has often very special characteristics (morphology, physiology, etc.) which are different from the adult phase.

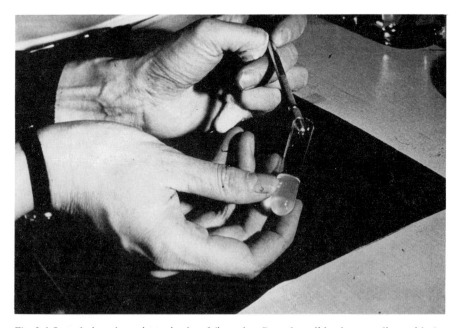

Fig. 2.4. Inoculation. A meristem is placed (inoculated) on the solid culture medium with the aid of an inoculation needle.

Laminar air-flow cabinet: Cabinet for inoculation which is kept sterile by a continuous non-turbulent flow of sterilized air.

Lateral: From the base of a leaf, bract or involucral bract.

Leaf primordium: Initial of a leaf.

Liquid media: Media without a solidifying agent such as agar.

Macro-elements: Group of essential elements such as N, P, K, Ca and Mg, normally required in relatively large quantities (inorganic nutrition of the plant).

Magnetic stirrer: Apparatus often consisting of a hot plate on which e.g. a beaker can be heated while a magnetic rod rotates inside.

Malt extract: Mixture of compounds from malt.

Medium: See nutrient medium.

Membrane filter: Aid in filter sterilization.

Mericlone: An orchid clone, originating from a meristem or other organs isolated in vitro.

Meristem: Collection (group) of dividing cells in the tip of a root, shoot (apical meristem), in the intercalary cambium of buds, leaves and flowers.

Meristemoid: Nodule of undifferentiated tissue from which new cells and/or adventitious structures (such as meristems) arise.

Micro-elements: Group of elements such as Fe, B, Zn, Mo, Mn, etc., important in relatively small quantities for inorganic nutrition of plants.

Micropropagation: Vegetative propagation of plants in vitro.

Monolayer: A single layer of cells growing on a surface.

Monoploid: A cell or individual which only has one genome ($2n=x$) present. A monoploid arises from a diploid ($2n=2x$) (Hermsen, 1977) and is the lowest number of chromosomes of a polyploid series (de Fossard, 1976).

Morphogenesis: The origin of form and, by implication, the differentiation of associated internal structural features (de Fossard, 1976).

Mutation: Genetic change.

Mycorrhiza: An association of a fungus with the root of a higher plant.

N: Denoted in the text by n, this is the number of chromosomes in the haploid phase (gametophyte).

Nucellar embryo: Embryo developed vegetatively from somatic tissue surrounding the embryo sac, rather than by fertilization of the egg cell.

Nutrient medium: Mixture of substances on/in which cells, tissues or organs can grow, with or without agar.

Organ: Part of a plant with a specific function, e.g. root, stem, leaf, flower, fruit, etc.

Organ culture: Culture of an organ in vitro in a way that allows development and/or preservation of the originally isolated organ.

Organ formation (organogenesis): Formation of a root, shoot, bud, flower, stem, etc.

Osmotic potential (value): Potential brought about by dissolving a substance e.g. in water; 1 mole of glucose per liter of water generates an osmotic potential of 22.4 atmospheres.

Parthenogenesis: Production of an embryo from a female gamete without the participation of a male gamete (de Fossard, 1976).

Pathogen-free: Plant, meristem, tissue, or cell, which is free of diseases (bacteria, fungi, viruses, etc.).

Petri dish: Flat, round dish made of glass or synthetic material with a cover.

pH: The negative logarithm of the concentration of hydrogen ions.

Plating efficiency: The percentage of cells planted which give rise to cell colonies.

Polyembryony: When two or more embryos are formed after fertilization (de Fossard, 1976).

Polyhaploid: This is an individual with half the number of chromosomes of a polyploid plant.

Polyploid: A plant is polyploid when $2n = 3x$ (triploid), $2n = 4x$ (tetraploid), etc. It has multiples of the base (lowest) number of chromosomes (x) (de Fossard, 1976; Hermsen, 1977).

Primary culture: Culture resulting from cells, tissues, or organs taken from an organism.

Primordium: Group of cells which give rise to an organ.

Protocorn: Tuberous-like structure formed when orchid seeds germinate or when meristem culture of orchids is applied.

Protoplast: Plant cell without a cell wall, produced by enzymatic degeneration of the cell wall.

Regulator: Substance regulating growth and development of plant cells, organs, etc.

Rejuvenation: Reversion from adult to juvenile.

Rotary shaker: Rotating machine on which e.g. Erlenmeyer flasks containing liquid nutrient medium can be shaken.

Scalpel: Small, sharp surgical knife.

Shoot tip: Apical (terminal) or lateral shoot meristem with a few leaf primordia or leaves.

Single-node culture: Culture of separate lateral buds, each carrying a piece of stem tissue.

Solid media: Nutrient media solidified e.g. with agar.

Somaclonal variation: Increase of genetic variability in higher plants which takes place in in vitro culture.

Somatic hybridization: Hybridization at the somatic level.

Sphaeroblast: Nodule of wood which can give rise to adventitious shoots with juvenile characteristics.

Stereomicroscope: See binocular.

Sterile: Medium or object with no perceptible or viable micro-organisms. Sterility tests are necessary for substantiation.

Sterilization: Procedure for the elimination of micro-organisms.

Sterilize: Elimination of micro-organisms, e.g. by chemicals, heat, irradiation or filtration.

Sterile room: Operation room for plants, inoculation room; at present replaced by laminar air-flow cabinets.

Steward bottle: Flask for the growth of cells and tissues in a liquid medium, in which they can be periodically submerged during rotation. Developed by Steward.

Subculture: Transplanting a cell, tissue or organ, etc. from one nutrient medium to another (Fig. 2.5).

Subculture number: The number of times cells, etc. have been subcultured i.e. transplanted from one culture vessel to another.

Suspension culture: A type of culture in which (single) cells and/or clumps of cells grow and multiply while suspended in a liquid medium.

Symbiosis: Two different organisms living together to the mutual advantage of both.

Test tube: Tube in which cells, tissues, etc. are cultured.

Fig. 2.5. For vegetative propagation of ferns in vitro, shoots are subcultured on a new (fresh) culture medium.

Thermolabile: Not heatproof, e.g. a substance which disintegrates upon heating.

Tissue culture: The culture of protoplasts, cells, tissues, organs, embryos or seeds in vitro.

Totipotency: Potential of cells or tissues to form all cell types and/or to regenerate a plant.

Transformation in vitro: The production, for whatever reason of hereditory changes by the growth of protoplasts, cells, tissues, etc.

Virus-free: Plant, cell, tissue or meristem which exhibits no viral symptoms or contains no identifiable virus-particles.

Vitamins: Group of organic compounds sometimes added to nutrient media (vitamin B_1, vitamin C, etc.).

Vitrification: Physiological disease, further explained in Section 6.4.2.

Vitro: Literally in glass, in a test tube, bottle, etc.

Vivo: In situ. In the intact plant growing in the greenhouse, the field, etc.

X: In the text x. In Chapter 23 on haploids this is often used. It denotes the smallest number (both structurally and in gene content) of different chromosomes which can function together as a single unit (Hermsen, 1977).

Yeast extract: Mixture of substances from yeast.

3. History

3.1. Introduction

The history of in vitro culture is illustrated in Fig. 3.1. In 1838 Schwann and Schleiden (cf. Gautheret, 1983) put forward the so-called totipotency theory, which states that cells are autonomic, and in principle, are capable of regenerating to give a complete plant. Their theory was in fact the foundation of plant cell and tissue culture. The first attempts, by Haberlandt in 1902, at plant tissue culture techniques failed. However, between 1907 and 1909, Harrison, Burrows and Carrel succeeded in culturing animal and human tissue in vitro. Although earlier workers had achieved in vitro culture of orchid seeds (seedlings), embryos and plant organs, in 1939 Nobécourt, Gautheret and White (cf. Street, 1973) succeeded in obtaining the first real plant tissue culture. After the 2nd World War development in this field was especially rapid and numerous results of importance for agriculture, forestry and horticulture have been published (Pierik, 1979; Bhojwani et al., 1986).

Plant tissue culture lagged behind animal and human tissue culture because of the late discovery of plant hormones (regulators). The first regulator to be discovered, the auxin IAA created great opportunities for the in vitro culture of plant tissues. The discovery of the regulator kinetin (a cytokinin) in 1955 was a further stimulus. Since that time tremendous developments have taken place, initially in France and U.S.A., but later in other countries. The substantial increase in the number of research workers in the last 5 years can be accounted for by the benefits which in vitro culture of higher plants offers the agriculturalists, plant breeders, botanists, molecular biologists, biochemists, plant pathologists, etc. Since plant tissue culture has such far reaching practical consequences for agriculture and plant breeding etc., the number of research workers will certainly continue to increase in the future.

For a detailed historical survey the reader is referred to the books and

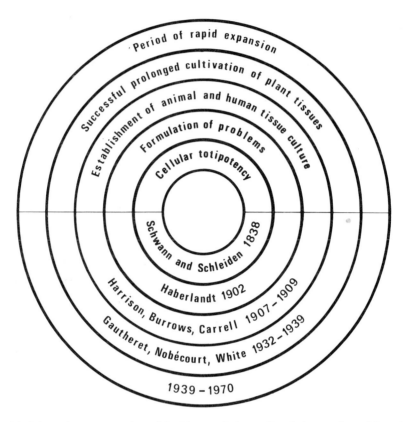

Fig. 3.1. Schematic representation of the history of plant cell and tissue culture. More than 100 years passed before the totipotency theory of Schwann and Schleiden (Gautheret, 1983) was substantiated.

articles by Gautheret (1959, 1964, 1983), Street (1973, 1974), and Gamborg and Wetter (1975). Only a few important facts and dates are mentioned below:

1892 Plants synthesize organ forming substances which are polarly distributed (Sachs).
1902 First attempt at plant tissue culture (Haberlandt).
1904 First attempt at embryo culture of selected crucifers (Hännig).
1909 Fusion of plant protoplasts, although the products failed to survive (Kuster).
1922 Asymbiotic germination of orchid seeds in vitro (Knudson).
1922 In vitro culture of root tips (Robbins).
1925 Embryo culture applied in interspecific crosses of *Linum* (Laibach).

22

1929 Embryo culture of *Linum* to avoid cross incompatibility (Laibach).

1934 In vitro culture of the cambium tissue of a few trees and shrubs failed to be sustained since auxin had not yet been discovered (Gautheret).

1934 Successful culture of tomato roots (White).

1936 Embryo culture of various gymnosperms (LaRue).

1939 Successful continuously growing callus culture (Gautheret, Nobécourt and White).

1940 In vitro culture of cambial tissues of *Ulmus* to study adventitious shoot formation (Gautheret).

1941 Coconut milk (containing a cell division factor) was for the first time used for the culture of *Datura* embryos (van Overbeek).

1941 In vitro culture of crown-gall tissues (Braun).

1944 First in vitro cultures of tobacco used to study adventitious shoot formation (Skoog).

1945 Cultivation of excised stem tips of *Asparagus* in vitro (Loo).

1946 First whole *Lupinus* and *Tropaeolum* plants from shoot tips (Ball).

1948 Formation of adventitious shoots and roots of tobacco determined by the ratio of auxin/adenin (Skoog and Tsui).

1950 Organs regenerated from callus tissue of *Sequoia sempervirens* (Ball).

1952 Virus-free dahlias obtained by meristem culture (Morel and Martin).

1952 First application of micro-grafting (Morel and Martin).

1953 Haploid callus of *Ginkgo biloba* produced from pollen (Tulecke).

1954 Monitoring of changes in karyology and in chromosome behaviour of endosperm cultures of maize (Strauss).

1954 First plant from a single cell (Muir et al.).

1955 Discovery of kinetin, a cell division hormone (Miller et al.).

1956 Realization of growth of cultures in multi-litre suspension systems to produce secondary products by Tulecke and Nickell (Staba, 1985).

1957 Discovery of the regulation of organ formation (roots and shoots) by changing the ratio of cytokinin/auxin (Skoog and Miller).

1958 Regeneration of somatic embryos in vitro from the nucellus of *Citrus ovules* (Maheshwari and Rangaswamy).

1958 Regeneration of pro-embryos from callus clumps and cell suspensions of *Daucus carota* (Reinert, Steward).

1959 Publication of the first extensive handbook on plant tissue culture (Gautheret).

1960 First successful test tube fertilization in *Papaver rhoeas* (Kanta).
1960 Enzymatic degradation of cell walls to obtain large numbers of protoplasts (Cocking).
1960 Vegetative propagation of orchids by meristem culture (Morel).
1960 Filtration of cell suspensions and isolation of single cells by plating (Bergmann).
1962 The development of the famous Murashige and Skoog medium (Murashige and Skoog).
1964 First haploid *Datura* plants produced from pollen grains (Guha and Maheshwari).
1964 Regeneration of roots and shoots on callus tissue of *Populus tremuloides* (Mathes).
1965 Induction of flowering in tobacco tissue in vitro (Aghion-Prat).
1965 Differentiation of tobacco plants from single isolated cells in micro-culture (Vasil and Hildebrandt).
1967 Flower induction in *Lunaria annua* by vernalization in vitro (Pierik).
1967 Haploid plants obtained from pollen grains of tobacco (Bourgin and Nitsch).
1969 Karyological analysis of plants regenerated from callus cultures of tobacco (Sacristan en Melchers).
1969 First successful isolation of protoplasts from a suspension culture of *Hapopappus gracilis* (Eriksson and Jonassen).
1970 Selection of biochemical mutants in vitro (Carlson).
1970 Embryo culture utilized in the production of monoploids in barley (Kasha and Kao).
1970 First achievement of protoplast fusion (Power et al.).
1971 First plants regenerated from protoplasts (Takebe et al.).
1972 Interspecific hybridization through protoplast fusion in two *Nicotiana* species (Carlson et al.).
1973 Cytokinin found capable of breaking dormancy in excised capitulum explants of *Gerbera* (Pierik et al.).
1974 Induction of axillary branching by cytokinin in excised *Gerbera* shoot tips (Murashige et al.).
1974 Regeneration of haploid *Petunia hybrida* plants from protoplasts (Binding).
1974 Fusion of haploid protoplasts found possible which gave rise to hybrids (Melchers and Labib).
1974 Biotransformation in plant tissue cultures (Reinhard).
1974 Discovery that the Ti-plasmid was the tumour inducing principle of *Agrobacterium* (Zaenen et al.; Larebeke et al.).
1975 Positive selection of maize callus cultures resistant to *Helminthosporium maydis* (Gengenbach en Green).

1976 Shoot initiation from cryo-preserved shoot apices of carnation (Seibert).

1976 Interspecific plant hybridization by protoplast fusion for *Petunia hybrida* and *Petunia parodii* (Power et al.).

1976 Octopine and nopaline synthesis and breakdown found to be genetically controlled by the Ti-plasmid of *Agrobacterium tumefaciens* (Bomhoff et al.).

1977 Successful integration of the Ti-plasmid DNA from *Agrobacterium tumefaciens* in plants (Chilton et al.).

1978 Somatic hybridization of tomato and potato (Melchers et al.).

1979 Co-cultivation procedure developed for transformation of plant protoplasts with *Agrobacterium* (Marton et al.).

1980 Use of immobilized whole cells for biotransformation of digitoxin into digoxin (Alfermann et al.).

1981 Introduction of the term somaclonal variation (Larkin and Scowcroft).

1981 Isolation of auxotrophs by large scale screening of cell colonies derived from haploid protoplasts of *Nicotiana plumbaginifolia* treated with mutagens (Siderov et al.).

1982 Protoplasts are able to incorporate naked DNA; transformation with isolated DNA is consequently possible (Krens et al.).

1982 Fusion of protoplasts by electrical stimulus (Zimmermann).

1983 Intergeneric cytoplasmic hybridization in radish and rape (Pelletier et al.).

1984 Transformation of plant cells with plasmid DNA (Paszkowski et al.).

1985 Infection and transformation of leaf discs with *Agrobacterium tumefaciens* and the regeneration of transformed plants (Horsch et al.).

3.2. Developments in the Netherlands and other countries

In vitro culture of higher plants in the Netherlands expanded rapidly between 1964 and 1987 from 10 to 98 active research workers. An overview is given in the table below of the in vitro research in the Netherlands (at universities, institutes and research stations) in the spring of 1985. They have been subdivided and listed under their relevant discipline.

Discipline	Number of academic research workers
Botany (anatomy, morphology and physiology)	23
Phytopathology (mycology, bacteriology, virology and nematology)	7
Plant breeding, genetics and molecular biology	42
Vegetative propagation	9
Pharmacology and secondary metabolites	17
Total	98

Research workers active in tissue culture are also to be found in the plant propagation and plant breeding industry, where the vegetative propagation of plants in vitro is of primary importance. The Table (see also Chapter 26) below shows the number of commercial tissue culture laboratories in the Netherlands, which propagate plants in vitro and the number of plants propagated in the years 1983–1986 (Pierik, 1986; unpublished results).

Numbers of plant produced	Number of commercial tissue culture laboratories			
	1983	1984	1985	1986
Less than 10,000	6	3	12	14
10,000– 100,000	9	15	14	18
100,000– 500,000	4	4	6	6
500,000–1,000,000	1	2	1	2
1,000,000–5,000,000	8	9	7	7
More than 5,000,000	0	0	2	3
Total	28	33	42	50

Number of plants propagated	1983	1984	1985	1986
Pot plants	15,243,327	15,428,130	17,412,586	19,822,274
Cut flowers	3,634,497	10,036,990	11,420,824	12,639,758
Bulbs and corms	590,392	1,458,019	5,358,740	8,085,920
Orchids	1,347,350	1,534,500	1,116,740	1,449,190
Misc. ornamentals	8,952	242,383	301,865	375,805
Agricultural crops	279,400	280,000	300,000	311,825
Vegetables	27,053	60,040	71,205	68,825
Total	21,130,771	29,040,062	35,981,960	42,753,600

Regrettably the exact numbers of laboratories and their production rates are not available for other countries. Sluis and Walker (1958) estimated the number of commercial tissue culture laboratories in the USA and Canada to be as high as 250, although only 5 to 10 individual laboratories in these two countries produce over 5 million plants per year. Jones (1986) estimated the production in 1985 of 18 major U.S. laboratories to be as high as 55–65 million plants.

During a congress in Tenerife in 1985 the author met several French tissue culture researchers who estimated the total in vitro production of cloned plants in France in 1985 as being 71 million.

 **B.V. CLEAN AIR TECHNIEK WOERDEN**

LAMINAR CROSSFLOW (CLEAN) BENCHES

TYPE CLF	**SERIES CLF**	**BAUREIHE CLF**

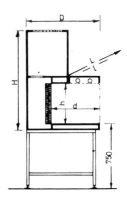

Changes without prior notice.

TYPE CLF/R	**SERIES CLF/R**	**BAUREIHE CLF/R**
Met voorrandafzuiging	With leading edge intake	Mit Vorderrandabsaugung

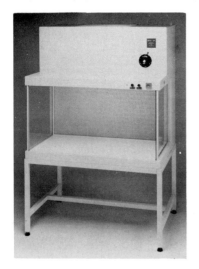

Best in Superb Clean Air Technology
Kuipersweg 37, Woerden – Phone 31.3480.11114 –
Telex 47219 – The Netherlands –
Pays-Bas – Die Niederlande

B.V. CLEAN AIR TECHNIEK WOERDEN

Quality from Holland

4. Types of culture

A plant consists of different organs, each being composed of different tissues, which in turn are made up of individual cells. If the cell walls of these cells are enzymatically digested, protoplasts are produced. Just as there are many different building materials within a plant, there are many different types of in vitro culture:

1. Culture of intact plants: a seed may be sown in vitro from which a seedling, and finally a plant develops e.g. orchids (Fig. 4.1.).
2. Embryo culture: here an isolated embryo is grown after removal of the seed coat.
3. Organ culture: an isolated organ is grown in vitro. Different types can be distinguished e.g. meristem culture, shoot-tip culture, root culture, anther culture, etc. Often a part (tissue mass, organ) which has been isolated from a plant is referred to as an explant and the culture of this is an explant culture.
4. Callus culture: if a differentiated tissue is isolated, allowed to de-differentiate in vitro and a so-called callus tissue produced, the process is termed callus culture.
5. Single-cell culture: the growing of individual cells which have been obtained from a tissue, callus or suspension culture, with the aid of enzymes or mechanically.
6. Protoplast culture: the culture of protoplasts obtained from cells by enzymatic digestion of the cell wall.

De Fossard (1977) differentiated between 3 types of in vitro culture in higher plants:

1. *Organized:* The culture of (almost) whole plants (embryo's, seeds, Fig. 4.1.) and organ culture are termed organized cultures; the characteristic organizational structure of a plant or the individual organ being maintained. It closely resembles in vivo vegetative propagation

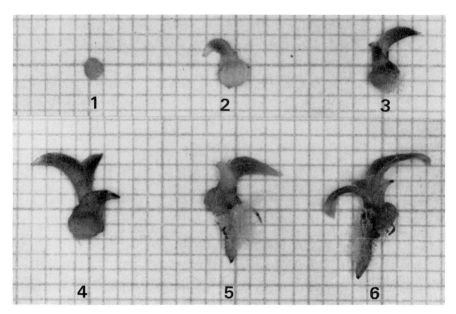

Fig. 4.1. Development of a *Paphiopedilum ciliolare* plant from an in vitro sown seed, as an example of organized growth and development. In 1 a protocorm is formed; in 2 this protocorm forms a leaf; in 3 two leaves are formed, while rhizoids are visible at the base of the protocorm; in 4 a third leaf is formed, and in 5 and 6 the first root arises (Pierik, unpublished). Photograph of 6 taken 20 weeks after isolation.

   by cuttings, division, runners and axillary buds or shoots. If the organizational structure is not broken down then progeny arise which are identical to the original plant material.

2. *Non-organized:* If cells and/or tissues are isolated from an organized part of a plant, de-differentiate and then are cultured, a non-organized growth (in the form of a callus tissue) results. If the callus disperses clumps of cells (aggregates) and/or single cells result (also referred to as suspension culture). Non-organized growth is mainly induced by the use of (very) high concentrations of auxin and/or cytokinin in the nutrient medium. The genetic stability of non-organized cultures is often low.

3. *Non-organized/organized:* This type of culture is intermediate between types one and two. Cells in an isolated organ or tissue, first de-differentiate and then form tissues or a layer of callus tissue, by division, from which organs (e.g. roots and/or shoots) or even whole individuals (pro-embryos or embryos) often rapidly develop. It must be taken into account that organized structures can develop from non-organized cultures either through special techniques or spontaneously. In all these cases the progeny are often not completely identical with the original plant material.

5. Laboratory equipment

5.1. Supplies

The necessary pieces of equipment required for setting up a professional tissue culture laboratory are listed below. For work on a small scale, simpler equipment can be used and for this the items marked with an asterisk (*) are not absolutely necessary. For example: washing up can be done by hand rather than with a washing-up machine, mixers are only really necessary if large quantities of media are required, distilled water can in some cases be replaced by de-ionized water, a domestic pressure cooker may be used instead of an autoclave. Only large scale laboratories have stores necessary for glassware, nutrient media etc.

Preparation of media

— Gas, water and electricity supplies and possibly compressed air* and vacuum lines*.
— Water heater (geyser).
— Different types of glassware.
— Hotplate with magnetic stirrer.
— Heating-jacket or hot-plate suitable for larger vessels; in practice a medium is often heated by the use of steam injection.
— Steamer for heating media*.
— Coarse balance for measuring in grams (accurate to 0.01 g).
— Sensitive balance for measuring in milligrams (accurate to 0.1 mg).
— Spatula for use during weighing.
— Mixer for large quantities of media*.
— Micro-wave oven for rapid heating of media and agar mixtures* (instead of a heating-jacket); the micro-wave oven can also be used for rapid thawing of frozen items.

- Millipore filtersystem*.
- Automatic dispensor*.
- pH-meter.
- Distillation apparatus*.
- De-ionizer.
- Wide range of (often expensive) chemicals.
- Kitchen timer for timing sterilizations.
- Metal racks for holding test tubes in the autoclave.
- Metal Petri dish holders for use in the autoclave*.
- Test tubes, flasks, plastic containers.
- Materials for closure of tubes and flasks (cotton plugs, aluminium foil, plastic film or metal caps, etc.).
- Autoclave* or pressure cooker.
- Storage space for chemicals, glassware etc.*.
- Sterile storage space for nutrient media, sterile water etc.*.
- Storage tank for distilled and/or de-ionized water.
- Drying and draining racks.
- Pipette washer* (acid-proof).

Isolation of cultures

- Laminar air-flow cabinet (sterile inoculation cabinet).
- Gass supply in inoculation cabinet*.
- Stereo-microscope*.
- Bunsenburner for flaming* (or spirit lamp).
- Dry sterilizer for knives, scalpels, forceps* (alternative to Bunsen-burner).
- Filterpaper (possible glass plates) for use during sterile cutting.
- Petri dishes for sub-culture.
- Centrifuge* (low speed).
- Hypochlorite for sterilization of plant material.
- 96% alcohol.
- Fire-proof storage space for alcohol.
- Vitafilm or similar material for wrapping boxes, racks, etc.

General equipment

- Dyring oven for drying glassware etc. after sterilization*.
- Automatic dishwasher*.
- Cleaning materials (brushes, etc.).
- Detergent.
- Trolley*.

— Refrigerator for storing chemicals and nutrient media.
— Small deepfreeze*.
— Acid proof bath containing chromic acid and sulphuric acid for cleaning strongly contaminated glassware*.

Culture room

— Temperature controlled (17–27 °C) with cooling as well as heating.
— Electricity supply essential for lighting, cooling and heating.
— Shelving for culture racks.
— Fluorescent tubes for lighting.
— Timer for regulating day-length.
— Alarm system in case of malfunction.
— Test tube racks.
— Table for making observations.
— Shaker of rotator*.

The above equipment will enable propagation work to be carried out, but the additional requirements for single cell culture and/or protoplast work have not been included.

Greenhouse space

For the growth of experimental material and further growth of in vitro produced plants.

From the overview given above it can be seen that the following working space is needed for an efficient tissue culture laboratory:

1. Greenhouse for cultivation and further growth of plant material.
2. Laboratory where the nutrient media can be prepared, preferably with a lot of cupboard space.
3. Storage space for all apparatus (glassware, chemicals, nutrient media, etc.).
4. Washing-up area for autoclave and dishwasher preferably a separate room from the laboratory mentioned under 2.
5. Inoculation room with laminar air-flow cabinets, preferably with sterile (filtered) air under positive pressure.
6. Room for balances (if possible balances in a separate part), refrigerator, deep freeze, etc.
7. Culture room(s).
8. Technicians room.

The mean area (expressed as percentages) allocated for separate functions in 8 commercial tissue culture laboratories in California are: media preparation 35%, inoculation room 23%, and culture rooms 42% (Marshall, 1977).

5.2. The laminar air-flow cabinet

Although the preparation and cutting of explants, the division (cutting) of calluses etc. takes place on a sterilized glass plate or between/on sterile filter paper, it is still necessary to carry out these tasks in a laminar air-flow cabinet (also called an inoculation cabinet). When carried out in a non-sterile room the number of infections is far greater, and if carried out on, or under sterile paper in non-sterile air (e.g. in the case of large pratical classes where the use of laminar air-flow cabinets would be too expensive) at least 10% infection must be expected. Research laboratories and commercial tissue culture laboratories always use laminar air-flow cabinets to limit the possibilities of infection.

A laminar air-flow cabinet (Fig. 5.1) is one in which the air is sucked from the outside, first being filtered through very fine filters before reach-

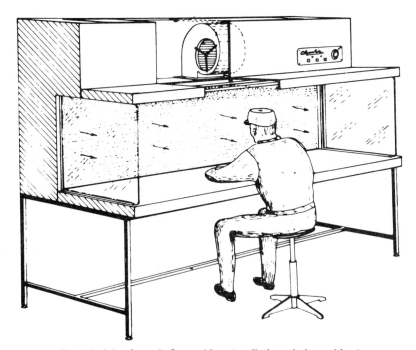

Fig. 5.1. A laminar air-flow cabinet (sterile inoculation cabinet).

34

ing the table top of the inoculation cabinet. This filtering system ensures that the air-flow over the table (this flow being laminar, giving the cabinet it's name) is completely sterile. Since there is a continuous air-flow through the inoculation cabinet, it is practically impossible that anything can pass from the outside room into the cabinet itself. When not in use the air-flow cabinet can be closed from the outside air with a plastic cover. The air-flow can be regulated and fluorescent tubes, for lighting, are fixed in the roof of the cabinet. A gas supply is available on the table top for use in flaming. However, flaming may be substituted by a spirit lamp or so-called dry-sterilization process. In this process the instrument to be sterilized is placed for a time in an apparatus containing heated small glass balls or in a so-called Bacti-Cinerator.

It is recommended that the cabinet is serviced by the factory once a year, and most laboratories find it best to take out a service contract with the supplier.

In modern laboratories the laminar air-flow cabinet is built in a special (relatively clean) isolation room, which is kept sterile and dust free by the use of filters. The room is pressurized so that non-sterile air from outside cannot enter.

The filters of the laminar air-flow cabinet should be regularly vacuumed and replaced annually (the coarse filters become particularly dirty). The floor of the room containing the cabinet should be decontaminated every day and only clean inside shoes and clean laboratory coats should be used. Visitors from outside are particularly a source of infection (via clothes and shoes). Other guidelines for use with an inoculation room are: to keep the room as clean as possible, not to bring any infected material inside, to remove tubes and containers suspected of being infected as soon as possible, to regularly remove rubbish in plastic bags, to regularly clean tables with 96% alcohol (not 70% alcohol which leaves water droplets behind) using a good quality cloth that leaves no pieces behind, to remove spore infected media, to regularly change instruments, not to allow any unnecessary object or person to enter, never to use the cabinet as a storage space, to regularly wash hands and arms with soap and then with 96% alcohol, to put used instruments directly into 96% alcohol or into the so-called dry-sterilizer, to switch on the laminar air-flow cabinet 15 minutes before use, to avoid interruption or reversal of the air-flow, to expose the most vital part of the cabinet to the direct sterile air-flow, and to avoid putting your head (hair) inside the cabinet itself. In some laboratories UV lamps are switched on at night to disinfect the air.

5.3. Sterilization of nutrient media

5.3.1. *Introduction*

Before seeds, parts of plants, organs, tissues, etc. are placed on a medium (inoculated), it must be sterilized i.e. made free of all micro-organisms. Sterilization can be carried out as follows:

1. Physical destruction of micro-organisms by dry hot air, steam or irradiation (UV light or gamma irradiation).
2. Chemical destruction of micro-organisms using sterilizing compounds (ethylene oxide, alcohol, hypochlorite, etc.) or antibiotics.
3. Physical removal of micro-organisms by filtration and/or washing.

Sterilization of nutrient media usually takes place in an autoclave (large pressure cooker), less often by filtration and seldom by irradiation. All methods along with their advantages and disadvantages can be found in a handbook (van Bragt et al., 1971).

The sterilized nutrient media, glassware, etc. should be stored in a sterile cupboard or metal box which has previously been disinfected with 96% alcohol.

5.3.2. *Autoclaving*

Most nutrient media are sterilized with the use of an autoclave. This (literally meaning container with a self sealing door) is an apparaturs for sterilization with steam. Providing exposure is sufficient, pressurized steam can destroy all micro-organisms. For example, nutrient media with a volume of up to 50 ml are kept for 20 min. at 121 °C in saturated steam under high pressure. The autoclave (Figs 5.2 and 5.3) is an expensive piece of apparatus, although it can be bought in different sizes and styles. There is a choice of horizontal (loading from the front, Fig. 5.2) and vertical (loading from above, Fig. 5.3). The horizontal autoclave is far easier to use, but is relatively expensive.

An autoclave has a temperature range of 115–135 °C. Good sterilization relies on: time, pressure, temperature and the volume of the object to be sterilized. Advantages of an autoclave are: speed, simplicity, the additional destruction of viruses and no adsorption (this occurs with filter sterilization). Disadvantages are: change in pH can result, components can separate out and chemical reactions can occur resulting in a loss of activity of media constituents.

Fig. 5.2. The horizontal autoclave (on the left) for sterilization of nutrient media.
Fig. 5.3. Two vertical autoclaves for the sterilization of nutrient media.

Guide-lines which must be followed when using an autoclave (N.B. the timing is started when the relevant temperature is reached):

1. Test tubes and flasks containing between 20–50 ml nutrient media: 20 min. at 121 °C.
2. Flasks containing 50–500 ml nutrient media: 25 min. at 121 °C.
3. Flasks containing 500–5,000 ml nutrient media: 35 min. at 121 °C.
4. Empty test tubes, flasks and filter paper: 30 min. at 130 °C.

For safety, when autoclaving bottles, they should not be too tightly packed and their tops should be loose.

Material that can be dry sterilized (such as test tubes, empty flasks and Petri dishes, paper, instruments, etc.) require 2–3 h dry sterilization at 160 °C.

As has already become clear, nutrient media and 'empty objects', such as glass, paper, etc. should be sterilized separately. The same holds true for large and small flasks. It must be realized that the heat penetration is very important in an autoclave; and large volumes must in principle be sterilized for longer periods, as the heat will take longer to penetrate than with smaller volumes.

Autoclaving breaks down the following (van Bragt et al., 1971):

— Saccharose: breaks down into fructose, glucose and sometimes levulose (an autoclaved medium with sucrose consists of different sugars).
— Colchicine (Griesbach, 1981).
— Zeatin (riboside).
— Gibberellic acid: 90% loss in reactivity (Fig. 5.4).
— Vitamin B_1 (pyrimidine and thiazol are formed), vitamin B_{12}, pantothenic acid, vitamin C.
— Antibiotics.
— Plant extracts (loss of activity).
— Enzymes.

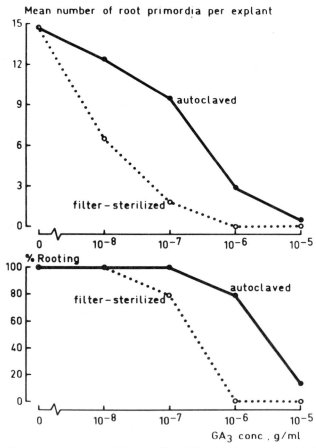

Fig. 5.4. The effect of autoclaved and filter-sterilized GA_3 solutions on auxin-induced adventitious root formation of in vitro cultivated bean *(Phaseolus vulgaris)* epicotyls (van Bragt et al., 1971).

Media to be used for protoplast culture (here there are far higher requirements than for media for explants and callus culture) are often filter-sterilized.

During and after autoclaving the following should be taken into account:

1. The pH of the media is lowered by 0.3–0.5 units.
2. Autoclaving at too high a temperature can caramelize sugars, which may then be toxic.
3. Autoclaving for too long can precipitate salts, and at the same time depolymerize the agar.
4. Care should be taken to use the correct duration and temperature (effective pressure). *Bacillus stearothermophilis* can be used as a biological indicator (to see if sterilization has been correctly carried out) as these are killed with 12–15 min. at 121 °C.
5. It must be realized that volatile substances can be destroyed by the use of an autoclave (e.g. ethrel, ethylene).
6. If nutrient media slopes are needed (e.g. for embryo culture and culture of orchid seeds), the test tubes must be placed on a slope to set after autoclaving (at a temperature of 45°–50 °C).
7. It is recommended that de-ionized water is used in the autoclave as tap water usually contains too much calcium which becomes precipitated on the bottom of the autoclave and the autoclave controls (water level and pressure controls).

In the last few years apparatus has become available in which a specific quantity (0.5–16 litres) of nutrient medium can be prepared and sterilized. During the sterilization the medium is mixed, helping the constituents to dissolve. This mixing also brings about more rapid heating and then cooling of the media at the end of the sterilization. After autoclaving thermolabile constituents can be added, after which it must be mixed again. The sterile medium can then be dispensed into sterile test tubes, flasks, etc. in the laminar air-flow cabinet. Automatic media preparation and sterilization units can be supplied e.g. by PBI International, Milano, Italy.

5.3.3. *Sterilization by irradiation*

Irradiation-sterilization of nutrient media (via gammarays) is hardly ever used in the growth of tissue cultures because it is extremely expensive when compared with the usual method of autoclaving. It appears however, that although gammaray-sterilization is an effective as with an autoclave, the resulting plant growth is significantly less on the media sterilized in this way. For further details see the handbook about sterilization by van Bragt et al. (1971).

When sterilizing plastic containers, boxes, tubes, etc. where the use of an autoclave is not possible, gammaray-sterilization is almost always used. The sterile autoclaved media are then dispensed into the sterilized containers in the laminar air-flow cabinet.

5.3.4. *Sterilization by filtration*

During filter-sterilization (solutions, liquid media, etc. pass through a membrane filter) all the particles, micro-organisms and viruses, which are bigger than the pore diameter of the filter used, are removed. The greatest advantage of this method is that thermolabile substances (broken down during autoclaving) can pass through the filter unchanged. Disadvantages can be: adsorption of substance into the filter, sometimes viruses particles pass through the filter, the procedure is time consuming and not as simple as autoclaving. All difficulties of filter-sterilization are found in the handbook by van Bragt et al., 1971.

Filter-sterilization is often used if a thermolabile substance X is needed in a nutrient medium A cellulose acetate, or cellulose nitrate filter is recommended with a pore diameter of 0.22 micron (μm). The basic medium without X is first autoclaved in a flask; while the medium is still liquid (approximately 45–50 °C), the liquid containing the substance X is injected with the help of a hypodermic syringe fitted with a membrane filter (carried out in the laminar air-flow cabinet). The medium containing X is then mixed and this complete nutrient media dispensed into previously sterilized tubes. It is also possible to filter a complete nutrient medium containing X.

Previously a vacuum was often used in conjunction with the filter-sterilization. The liquid was filtered, under vacuum, through a membrane into a Buchner flask, the vacuum being produced by an aspirator pump. Nowadays the solution to be filtered is usually forced through a filter by pressure, either with air pressure or that produced by a syringe. Obviously the parts of the filter (holder, filter, needle, etc.) must be sterilized before use, either with an autoclave or with 96% alcohol.

Problems can arise with filter-sterilization such as e.g. if a final concentration of substance X of 10 mg l^{-1} is needed (from a 1,000 × concentrated stock solution) if X is not very soluble in water. In such a case 1 ml of the filter-sterilized solution, containing 10,000 mg l^{-1} of substance X must be added to 1,000 ml of medium. This being impossible (due to the insoluble nature of X in water) then 10 mg of X can be weighed, sterilized in ether and then, after the ether has evaporated, added to the sterile medium.

5.3.5. *Sterilization at school and in the home*

An autoclave is normally too expensive to use at school or at home, but a pressure cooker, which is relatively inexpensive may be used instead. Nutrient media are best sterilized for 30 min. in a pressure cooker (Fig. 5.5) and large volumes e.g. flasks containing water or packs of filter paper for 60–70 min.

Fig. 5.5. The pressure cooker used for sterilization on a small scale at school or in the home by the amateur.

5.4. Preparation room

A well equiped laboratory needs a preparation room where the following should be available: autoclave, dishwasher, water de-ionizer, distilled water apparatus, washing-up area (with hot and cold running water). It is a great advantage if all these 'wet' procedures can be localized in one room. A dishwasher is strongly recommended since it is very time consuming to wash all the glassware by hand.

5.5. Cleaning glassware

As soon as glassware has been used it should be placed in warm water, especially when using solid nutrient media, since if agar is allowed to dry it sticks to the sides of the flasks and beakers and is much more difficult to remove. Cleaning should be done as soon as possible to prevent the media sticking to the sides of the containers, etc. After washing (preferably with warm water and soap), using a brush, the glassware should be rinsed, first with tap-water and then with distilled or de-ionized water, and finally checked to make sure it is completely clean. The glassware should be dried as soon as possible to reduce the chances of infections later, since micro-organisms need water to grow and reproduce.

Test tubes are cleaned as follows: firstly the stopper is removed and the agar and plant material taken out with a large pair of forceps. The test tubes are then washed under hot running water, using a brush which fits tightly inside and so cleans the sides. It is obviously much easier to use a dishwasher, the last rinse being with de-ionized water. It is best to sterilize the clean glassware before they are required again, drying them quickly and storing them in sterile conditions (upside down in a clean, dustfree cupboard).

5.6. Robots in the plant tissue culture laboratory

In a recent review article de Bry (1986) has answered the question, 'Why not robots in a plant tissue culture laboratory?'. He describes robots in general (history, development and classification), their general structure and operating systems, type of sensors, robot vision and also computer and artificial intelligence. He also describes possible applications (agricultural uses, laboratory automation and use in cell and tissue culture), the robots potential effects (cost comparisons, attitudes of researches to their use), and new challenges. This article contains much information and is worthwhile reading for those interested in the potential of robots for eliminating unpleasant work in plant tissue culture laboratories, particularly in plant propagation, and for reducing labour costs.

An intelligent robotic system for in vitro plantlet production was described by Deleplanque et al. (1985); this system locates plantlets on their medium, picks them up and places them in containers.

6. Preparation and composition of nutrient media

6.1. Introduction

The in vitro growth and development of a plant is determined by a number of complex factors:

1. The genetic make-up of the plant.
2. Nutrients: water, macro- and micro-elements, and sugars.
3. Physical growth factors: light, temperature, pH, O_2 and CO_2 concentrations.
4. Some organic substances: regulators, vitamins, etc.

The genetic make-up is a decisive factor at every stage in the plant, it determines, for example, if a plant is a monocotyledon or a dicotyledon, which leaf form it has, which temperature is optimal for growth and flowering, the flower form and colour, if a plant forms parthenocarpic fruits (without seeds) etc. The expression of the genetic make-up also depends on physical and chemical conditions which we also have to create in vitro.

Nutrients are essential for the growth and development of the plant: without water and mineral nutrients a plant cannot live in vitro or in vivo. Sugars must also be added to the culture medium, since plants (or parts of plants) in this condition, are not completely autotrophic.

The importance of physical factors in growth and development in vivo is just as applicable in vitro. These factors have an effect in all sorts of processes: water uptake, evaporation, photosynthesis, respiration, growth, flowering, fruit set, etc.

The fourth type of important factor is a group of organic substances to which regulators belong. Regulators which are only needed in very small concentrations, regulate, for example, the distribution of all sorts of substances within the plant and are therefore responsible for cell division, cell growth etc. Regulators, particularly auxin and cytokinin, regulate the development of organs (regeneration) on parts of plants (explants) grown in vitro. They are also very important in the development of (pro)embryos in suspension culture.

It is therefore obvious that the regulation of growth and development of a plant is a complicated process, which is dependant on the genetic make-up and also the environment. When preparing a nutrient medium so called 'interactions' should be borne in mind:

1. Regulator-temperature. Changes in organ formation can be based on the availability (endogenous or exogenous) of auxin and/or cytokinin. The endogenous auxin and/or cytokinin concentrations are sometimes strongly influenced by temperature.
2. Light/dark-auxin. Indoleacetic acid (IAA) is broken down more in light than in dark.
3. Genotype-regulator. Some genotypes require high and some low concentrations of regulators for growth and development. The potential for biosynthesis of a particular hormone is often genetically controlled.
4. Sugar-light. Sometimes light and sugar can (temporarily) replace each other. The replacement is also dependent on the photosynthetic capacity in vitro.

An explant, isolated in vitro and induced to grow and develop needs many substances that are schematically represented in Fig. 6.1. It can be seen from this figure that some of the components are also necessary for in vivo development (water, macro- and micro-elements). The organic substances and strange undefined mixtures of substances that a plant needs in vitro are not necessary in vivo, in other words a plant in vitro is heterotrophic. In any case it is definite that water, macro- and micro-elements, sugar, as a carbon source, and often two groups of regulators (auxins and cytokinins) are needed in vitro. If one of these components is lacking then normally no growth and development take place and the isolated organ or tissue dies. Sometimes the isolated tissue is able to biosynthesize a particular compound and so supply its own needs; if a plant in vitro has, for example, no need of auxin, then it is called an auxin autotroph.

As is shown in Fig. 6.1 sometimes undefined mixtures of substances are added to induce growth; the addition of these mixtures is usual when the researcher does not know the nutritional and hormonal requirements. Nowadays nutrient media are almost always made up of a mixture of synthetic substances. It is obvious that finding a balanced nutrient me-

Nutritional and hormonal requirements of plant tissue and organ cultures				
Water				pH
Organic substances	Macro Micro elements			
Sugars Amino acids Vitamins Regulators ⎰ Auxins Cytokinins Gibberellins Abscisic acid Ethylene	N P K Ca Mg S	Fe Co Zn Ni B Al Mn Mo Cu I		
Undefined mixtures of substances	Yeast extract Coconut milk Plant extracts Casein hydrolysate Pepton and trypton			

Fig. 6.1. Complex of substances often added to nutrient media to induce growth and development. The complex consists of water, organic compounds (on the left), inorganic compounds (on the right) and a group of undefined mixtures of substances (below).

dium that is suitable for the growth of tissues from a particular plant type (unknown in the literature) will require a great deal of time. If a mixture of about 20 different compounds is needed then it is often a question of trial and error to find the correct proportions. There are of course many papers (Gautheret, 1959; Gamborg et al., 1976) on the general composition of nutrient media and these can often be a great help.

If no gelling agent, such as agar is added to the medium it is known as a liquid rather than a solid medium. The use of agar is dealt with in Section 6.4.2.

For someone with little experience, it is often difficult to choose a medium. Which medium should be used if there is no literature available? De Fossard (1976) suggested that a so-called broad-spectrum experiment should be carried out. In this case he arrived at the conclusion that 4 (groups of) components were often vital in a nutrient medium:

sugar, macro-salts, auxins and cytokinins. Since the requirements of the experimental plant are not known in advance, these 4 (groups of) compounds are varied in 3 concentrations (low, medium, high):

1. Sugar (saccharose): 1–2–4%.
2. Macro-salts (according to Murashige and Skoog, 1962): $\frac{1}{4}$–$\frac{1}{2}$–full strenght.
3. Auxin (e.g. IBA): 0.01–0.5–5.0 mg l^{-1}.
4. Cytokinin (e.g. BA): 0.01–0.5–5.0 mg l^{-1}.

Since each of the 4 compounds is tested in 3 concentrations, we finally have $3^4 = 81$ combinations from which the optimal for growth is chosen. It is then necessary to carry out a more crucial experiment; if for example from the first experimental series the auxin concentration 0.5 mg l^{-1} is optimal, then the following auxin series should be tested 0.05–0.1–0.3–0.5–0.8–1.0–1.5 mg l^{-1}.

Another experimental approach to large scale testing of culture media variation is the so-called MDA (multiple-drop-assay) technique. This technique is particularly applied to protoplasts of *Nicotiana* and *Petunia* species and to cereal mesophyl protoplasts (Harms, 1984). This technique uses droplets of 40 µl as the experimental unit.

6.2. Glassware and plastics

In vitro culture can in principle be carried out in glass as well as in plastic containers of various shapes and sizes. Glass has the advantage that it is more durable and can be autoclaved. Plastic normally has only a limit life (inexpensive plastic is usually only used once) and is not always suitable for use in the autoclave. Plastic that can be used in an autoclave is much more expensive. The durability depends on its resistance to heat (autoclave, dishwasher) and to detergent. Plastic can also have the disadvantage that it gives off ethylene which can have damaging effects if it accumulates.

There is a wide range of types of glass which can have important physiological consequences. In research it is advisable to use Pyrex or a similar borosilicate glass. The inexpensive poor quality glass can, for example, give off toxic cations like sodium, lead and arsenic into the medium. Pyrex is definitely recommended for work with protoplasts, single cell culture and meristem culture.

In practical horticulture it has been shown that it is not necessary to use expensive Pyrex glassware. Inexpensive glassware can often be used without any damage. This is important economically. As well as inexpensive glassware a great deal of plastic is used. Since plastic containers are often sealed too tightly, care should be taken that there is no excessive accumulation of ethylene or CO_2.

The following glassware is used for tissue culture:
1. Test tubes (sometimes with screwtops).
2. Petri dishes.
3. Erlenmeyer flasks (for both solid and liquid media).
4. The so-called 'Steward flask' (see Fig. 6.2).
5. In practical horticulture even milk-bottles, juice-bottles and preserving jars may be used.

Fig. 6.2. The so-called Steward bottle with 'nipples'. Cells and cell clumps are periodically submerged as rotation occurs.

In commercial tissue culture laboratories, particularly for propagation, all sorts of plastic tubes, boxes, tubs, etc. are used. When the plastic is not suitable for autoclaving it is packed in plastic bags (without media), sealed and sterilized by gamma rays. Previously sterilized media can then be dispensed into these in a laminar air-flow cabinet.

The shape and size of the glassware (plastic) can have important consequences. As an extreme example, we can compare a test tube with a Petri dish. The test tube has a relatively large volume and a very small surface area, resulting in little aeration and drying out, and so a low chance of contamination. On the other hand, the Petri dish has a small volume and a relatively large surface area; aeration is good, drying out is far more likely and the chance of contamination is relatively high. It is therefore more important to seal Petri dishes (with parafilm, etc.) than test tubes.

In general, it is true that cultures in larger flasks grow and develop (propagate) better than in small flasks. This is probably due to:

1. The volume is larger and the evolved toxic gases (CO_2, ethylene, etc.) are more diluted.
2. The larger available amount of nutrient media per unit.

Pierik and Bouter (unpublished) have recently established that *Phalaenopsis* seedlings appear to grow much better in glass pots than in test tubes.

6.3. Preparation

A large number of substances and even sometimes mixtures of substances are added to the nutrient media. It is advisable to use a spatula for each substance to avoid contamination. Unused compounds should never be returned to the storage jars. The concentration of a particular substance can be given in different ways. The following terms are general:

— Volume percentage: Used for coconut milk, tomato juice etc.; 5% coconut milk represents 50 ml added to 950 ml water.
— Weight percentage: Used when dealing with agar and sugars; 2% represents 20 g of additive per 1000 g (litre) of nutrient media.
— Molar: 0.01 M represents 1/100 mol per litre (1 mol is the same number of grams as the molecular weight). This is often used with regulators.
— Milligram per litre ($mg\,l^{-1}$): 10^{-7} represents $0.1\ mg\,l^{-1}$, 10^{-6} represents $1\ mg\,l^{-1}$. This is also used a great deal with regulators.

— Microgram per litre: $1\ \mu g\ l^{-1}$ represents $0.001\ mg\ l^{-1}$.
— Parts per million (ppm): 1 part per million represents $1\ mg\ l^{-1}$.

In relation to these, the concentration of a particular substance in a medium can be considered. This can, in principle be given in two ways:

1. *Units in weight:* e.g. $mg\ l^{-1}$ or $g\ l^{-1}$. In the literature a concentration of $1\ g\ l^{-1}$ may be given as 10^{-3} and a concentration of $1\ mg\ l^{-1}$ as 10^{-6}. This can also be given as ppm: 1 part per million represents 10^{-6} or $1\ mg\ l^{-1}$.
   Plant physiologists find the use of units of weight unacceptable since it is incorrect to compare the physiological activity of $1\ mg\ l^{-1}$ IAA and $1\ mg\ l^{-1}$ IBA. The correct comparison would be $1\ \mu M$ IAA with $1\ \mu M$ IBA since $1\ \mu M$ IAA contains the same number of molecules as $1\ \mu M$ IBA, not the case with $1\ mg\ l^{-1}$ IAA and $1\ mg\ l^{-1}$ IBA. This arises because the molecular weight of IAA and IBA are different, and with most substances, certainly with regulators, the concentration should in principle be given as molar concentration.
2. *Molar concentration* (M), millimolar (mM) or micromolar (μM): A molar solution (M) contains the same number of grams of the substance as is given by the molecular weight; 1 millimolar (mM) is 10^{-3} M and 1 micromolar (μM) is 10^{-6} M or 10^{-3} mM.

Since the concentration in Molar is being used more and more in plant tissue culture a few examples are given below.

Molecular weight of auxin IAA $= 175.18$
a 1 M IAA solution consists of 175.18 g per litre
a 1 mM IAA solution consists of 0.17518 g per litre $= 175.18\ mg\ l^{-1}$
a $1\ \mu M$ IAA solution consists of 0.00017518 gram per litre $= 0.17518\ mg\ l^{-1}$

Conversion from $mg\ l^{-1}$ to (milli)molar and back can be shown as follows for $CaCl_2 . 2\,H_2O$. The molecular weight of $CaCl_2 . 2\,H_2O$ is:
$40.08 + 2 \times 35.453 + 4 \times 1.008 + 2 \times 16.00 = 147.018$ (the atomic weights of Ca, Cl, H and O being 40.08, 35.453, 1.008 and 16.000 respectively). The number of mmol $CaCl_2 . 2\,H_2O$ per litre can be found by dividing the number of mg $CaCl_2 . 2\,H_2O$ per litre by the molecular weight of $CaCl_2 . 2\,H_2O$ (inclusive of water of crystallization) i.e. $440\ mg\ l^{-1}$ $CaCl_2 . 2\,H_2O = 440/147.918 = 2.99\ mM\ CaCl_2 . 2\,H_2O$. Alternatively:

$2.99 \text{ mM } CaCl_2.2H_2O = 2.99 \times 147.018 = 440 \text{ mg l}^{-1} CaCl_2.2H_2O.$
From the above calculation it can be seen that it is unnecessary to know the number of molecules of water of crystallization when using molar or millimolar concentrations.

To compare 2 different media fairly it is necessary to calculate the number of mg l^{-1} in mM. This is shown as follows:

Medium X comprises of	Medium Y comprises of
1650 mg l^{-1} NH$_4$NO$_3$	500 mg l^{-1} NH$_4$NO$_3$
1900 mg l^{-1} KNO$_3$	500 mg l^{-1} KNO$_3$
440 mg l^{-1} CaCl$_2$.2H$_2$O	347 mg l^{-1} Ca(NO$_3$)$_2$

If we calculate this in mmol this means per litre:

20.6 mM NH$_4$NO$_3$	6.25 mM NH$_4$NO$_3$
18.8 mM KNO$_3$	4.95 mM KNO$_3$
2.99 mM CaCl$_2$.2 H$_2$O	2.11 mM Ca(NO$_3$)$_2$

Calculated for the different ions:

20.6 mM NH$_4^+$	6.25 mM NH$_4^+$
39.4 mM NO$_3^-$	15.42 mM NO$_3^-$
18.8 mM K$^+$	4.95 mM K$^+$
2.99 mM Ca^{2+}	2.11 mM Ca^{2+}
5.98 mM Cl$^-$	0 mM Cl$^-$
60.0 mM N total	21.67 mM N total

From the calculations above it can be seen that the medium X has approximately 3 times as much nitrogen (NO$_3^-$ and NH$_4^+$) as the medium Y.

Nutrient media should be prepared as follows. A flask with distilled water is put in the heating unit, which is switched on. Approximately 550 ml of water should be warmed, if 600 ml of nutrient media is required. The necessary weight of sugar is added (while this is being carried out, take the flask out of the warming jacket). Add the macro- and micro-salts and then the auxins and/or cytokinins or other components. Mix the media well until the substances have dissolved and bring the solution to the boil. Add agar (using a funnel) as needed being careful at each addition that lumps and frothing are avoided. Alternatively the agar can be mixed with cold water in a beaker and this agar mixture added. The agar should not be added if the water is boiling as this can result in overcooking of the medium with detrimental effects. Finally water is added to make the volume to 600 ml. The pH can be measured with litmus paper or a pH meter and adjusted if necessary. The pH should be

6.0 when all the components have been added and the volume made up to 600 ml. The test tubes, flasks etc. can now be filled with medium. Test tubes and Erlenmeyer flasks are usually filled to one third capacity. A Büchner flask is used to dispense the nutrient media into test tubes to prevent media coming into contact with the walls and the outsides of the tubes. The following points are important for media preparation:

1. Where ever possible use analytical chemicals and weigh these out, using a clean spatula, on a sufficiently sensitive balance.
2. Sometimes use ready made media for economical reasons (see Section 6.5).
3. Agar and sugar can be weighed as needed, but macro- and microelements should be used from stock solutions. Stock solutions of regulators and vitamins are often prepared fresh.
4. Stock solutions which are stored at low temperatures sometimes acquire a precipitate. This can cause problems since if it is re-dissolved by boiling, water can be lost by evaporation.
5. If there is strong growth for a few weeks, the nutrients may be exhausted and subculture should be carried out.
6. Steaming the media (instead of boiling) to dissolve the agar is safer and causes less chemical decomposition.
7. Thermolabile substances (solutions) should be added by filter-sterilization after the basic media have been autoclaved and cooled to 45–50 °C.

Which media to use is difficult to answer since it depends on so many factors:

1. The experimental plant. Some species for example are very sensitive to salt while others can tolerate a high salt concentration. Some species react to the addition of vitamin B_1 and others don't. The need for regulators, especially auxins and cytokinins is very variable etc.
2. The age of the plant: juvenile tissues can regenerate roots without auxin but the adult tissues require the presence of auxin.
3. The age of the organ. Young organs (actively cell dividing) have different hormonal requirements than other tissues.
4. The type of organ cultured: if roots are cultured then a pre-requisite for vitamin B_1 is exhibited.
5. The need for regulators with suspension culture is less if the callus is grown for a longer period of time.
6. Every process carried out in vitro has it's own requirements. For example: adventitious roots often only arise after the addition of auxin, while adventitious shoots can arise after the addition of cytokinin.

When the nutrient requirements of a plant are unknown, the Mura-shige and Skoog (1962) macro- and micro-element mixture can be chosen, providing that the plant is not sensitive to high salt concentrations. If the plants are slightly sensitive to salt the macro- and micro-element mixture of Heller (1953) can be chosen and the Knop (1884) macro-element mixture if the species is very sensitive to salt. The usual form of sugar is 2–3% saccharose (sucrose). If a solid medium is needed then finally 0.7% Difco Bacto agar is added as a gelling agent. The addition of auxins and cytokinins is very difficult to prescribe and will be dealt with in detail in Section 6.4.7. The addition of vitamins and miscellaneous compounds will be dealt with in Sections 6.4.8 and 6.4.9 respectively.

6.4. Composition

6.4.1. *Water*

Great attention should be paid to the quality of the water since 95% of a nutrient medium consists of water. For research purposes it is recommended that water is used that has been Pyrexglass distilled and for research with protoplasts, cells and meristems, water which has been double distilled. To make sure of a good quality distilled water, which is efficiently produced, the distillation apparatus should be regularly cleaned to avoid furring and rinsed before use with de-ionized water. To avoid particles from the de-ionized water being left in the distillation apparatus a filter can be placed between the de-ionizer and the still.

In the last few years water has also been used that has been purified by reverse osmosis; this purification is then combined with other methods of purification (de-ionization, distillation, filtration).

Distilled water is best stored in polythene bottles, since glass often contains traces of lead, sodium, arsenic, which can be released into the water. If glass must be used for storing water then it should be Pyrexglass. Bottles which are to be used for water storage should be especially well cleaned.

Tap-water should not be used by those at school and by amateurs in the home. Where distilled water is unavailable then de-ionized water can be used, but it must be borne in mind that organic contaminants and even micro-organisms can be present. However, de-ionized water is often used for elementary practicals in which large explants are grown in vitro.

6.4.2. *Agar*

Agar is a seaweed derivative, obtained in pellet form, which can be used as a gelling agent in most nutrient media. Although a natural plant product, it is purified and washed by the manufacturers so that there are almost no toxic materials present. Agar is a polysaccharide with a high molecular mass that has the capability of gelling media. It is easily the most expensive component of solid nutrient media.

Solubilized agar forms a gel that can bind water (the higher the agar concentration, the stronger the water is bound), and adsorb compounds. If the agar concentration is increased, it is more difficult from an explant to make contact with the medium, which limits the uptake of compounds.

Plant tissue culturists often use Difco Bacto agar at a concentration of 0.6–0.8%, although other forms of agar (Gibco Phytagar, Flow agar etc.) are also becoming well known. Sometimes Difco Noble agar or purified Difco agar is used, although in most cases this is not necessary. When culturing protoplasts or single cells a very pure expensive agar, such as agarose (SeaPrep or SeaPlaque from Marine Colloids) is often used.

If different kinds of agar (different manufacturers) are compared, few significant differences are found. Roberts et al. (1984) compared the effect of six different agars on tracheary element differentiation in *Lactuca* explants. There were no significant differences in tracheary count, when using Difco Bacto agar, Gibco Phytagar and Flow agar, but when KC-TC agar was used more tracheary elements were produced. Significantly smaller numbers of tracheary elements were differentiated in the presence of the two FMC agaroses (SeaPrep and SeaPlaque) when compared to the other agars tested.

It must be remembered that agar also contains organic and inorganic contaminants (Romberger and Tabor, 1971), as can be seen from the table below for Difco-agars (supplied by Difco in 1984).

	Bacto-Agar	Noble-Agar	Purified Agar
Ash	4.50%	2.60%	1.75%
Calcium	0.13%	0.23%	0.27%
Barium	0.01%	0.01%	0.01%
Silica	0.19%	0.26%	0.09%
Chloride	0.43%	0.18%	0.13%
Sulphate	2.54%	1.90%	1.32%
Nitrogen	0.17%	0.10%	0.14%

A manufacturers analysis shows that Difco Bacto agar also contains 0.0–0.5 cadmium; 0.0–0.1 chromium; 0.5–1.5 copper; 1.5–5.0 iron; 0.0–0.5 lead; 210.0–430.0 magnesium; 0.1–0.5 manganese and 5.0–10.0 zinc (amounts given in ppm).

55

The usual concentration for agar is 0.6–0.8%. If a lower concentration (0.4%) is used then the nutrient medium remains sloppy, especially when the pH is also low. If a high concentration (1.0%) is chosen, then the nutrient medium is very solid, making inoculation difficult. If 0.6% is used and the medium remains sloppy then the pH should be corrected; if the pH is lower than 4.5–4.8 a medium with 0.6% agar does not gel properly.

In vitro growth may be adversely affected if the agar concentration is too high (Stolz, 1967). The type of agar also affects growth and development (Debergh, 1983). The better quality and purer agars gel better than the others available.

The following alternatives to agar are available:

1. Synthetic polymers (Gas et al., 1971) such as biogel P200 (polyacrylamide pellets).
2. Alginate can be used for plant protoplast culture (Mbanaso and Roscoe, 1982).
3. A stabilizing substrate derived from cellulose crystallite aggregates (CCA) was described by George et al. (1983) which is particularly useful in the rooting phase with cauliflower cultures.
4. Gelrite (tradename of Merck, USA, a division of Kelco), which is a highly purified gelling agent. It is a natural anionic heteropolysaccharide that forms rigid, brittle agar-like gels in the presence of soluble salts. Gelrite is polysaccharide comprised of glucuronic acid, rhamnose, glucose and O-acetyl moieties. The gel strength of Gelrite is highly dependent on the type of salt added; divalent cations such as magnesium (0.1% $MgSO_4 . 7 H_2O$) or calcium have a far greater effect on gel strength than monovalent ions such as sodium or potassium. Gelrite requires both a heating cycle and the presence of cations for gelation to take place.

   Gelrite may be used at approximately half the concentration of agar; for plant tissue culture 0.2% Gelrite is advised. Gelrite gels are remarkable clear in comparison to those formed by agar and also set more rapidly. According to the manufacturers, Gelrite contains no contaminating materials (e.g. phenolic compounds), which can be toxic. For further details see Halquist et al. (1983).
5. Cooke (1977a) also used an agar substitute (a starch polymer) for shoot-tip culture of *Nephrolepis exaltata*. This agar-substitute (concentration 12 g l⁻) is soluble in water without heating.
6. Use can be made of a liquid medium without agar, using clean foamplastic (Fig. 6.3) glasswool or rockwool as support.
7. Growth on a bridge of filter paper, which is hung in a liquid medium (Fig. 6.4). This system was developed by Heller (1953).

Fig. 6.3. In vitro culture of stem explants of *Rhododendron* 'Catawbiense album' in a liquid medium. The explants are apolarly (basal ends up) placed in a hole made in the foam plastic. Callus and adventitious roots are formed at the basal ends of the explants. Photograph taken 8 weeks after isolation.

8. Growth on a liquid medium, containing glass beads which can anchor explants.
9. Viscose sponge underneath filter paper can be used as a carrier for a liquid medium instead of agar. The advantages of viscose are (Wersuhn and Fritze, 1985):
    - It can be used repeatedly.
    - It facilitates large screening.
    - Subcultures can be carried out without changing the container.
    - Plants can be more easily transferred to soil.
    - Smaller amount of chemicals are required.

Fig. 6.4. Culture of a protocorm of the orchid *Cattleya aurantiaca* on a bridge of filter paper which is kept moist by the liquid medium in the test tube. The upper part of the filter paper lies just above the liquid medium.

10. Growth on a medium, without the use of any means of anchoring the material (as is the case for 1–9). If this system is used the cells, tissues, etc. are immersed and require good aeration. This is achieved with the use of shaker and is described in greater detail in Chapter 12.

In recent years the use of a so-called 2-phase medium has become quite common (a solid and a liquid phase), in which a high multiplication rate is combined with lower sensitivity for vitrification. The 2-phase medium is made as follows: firstly a solid medium is made into which explants are inoculated, and then a layer of liquid medium is added. The explants are then partly in solid agar, partly in liquid medium and partly in air. The composition of the liquid phase of the medium can easily be changed, e.g. to induce shoot elongation.

The decision whether to use a solid or liquid medium depends on the following:

1. Is the experimental plant suitable for growth in a liquid medium? While some plants do not grow well in liquid medium others prefer it (e.g. *Bromeliaceae*).
2. If it is possible to grow plants in a liquid medium the next important consideration is aeration. Most plants, especially if grown submerged need a good oxygen supply. This is possible by growing the explant half-submerged or if total submersion is necessary, by the use of a shaker. If a shaker is used in the growth of cells, tissues etc. then any possible damage caused by the agitation should be borne in mind. On the other hand growth and development in a liquid medium is often very good, since if the plants are submerged they can readily take up nutrients, regulators etc. on all sides; in contrast to growth on agar where there is only basal contact. Growth on liquid medium also means that any exudate from the explants is diluted more readily than on agar where local accumulations may occur.
3. Growth and organ differentiation can be completely different in liquid than on solid media. If *Anthurium andreanum* callus is grown in a liquid medium, then almost no adventitious shoots are formed, in contrast to their formation on a solid medium. If shoots of bromeliads or *Nepenthes* are grown in vitro then axillary shoot formation is much better in a liquid medium.
4. Liquid, as well as solid media can cause a physiological condition known as vitrification (Anonymus, 1978a; Debergh, 1983; Ziv and Halevy, 1983; Ziv, 1986).

In the literature vitrification is termed: translucency, hyperhydration, hyperhydric transformation, glauciness, waterlogging, glassiness. Vitrification is especially common if the plant has too much water available, this being usually the case in liquid media, or if the medium has a low agar concentration. It is described in *Malus* and *Prunus* (Anonymus, 1978a) in carnation (Ziv and Halevy, 1983) and artichoke (Debergh et al., 1981).

The occurrence and degree of vitrification is influenced by many complex factors:

1. Raising the agar and/or sugar concentration often lessens the vitrification in the cases of carnation and artichoke.

59

2. It may be prevented in some cases, by better gaseous exchange in the culture vessels.
3. It is associated with high cytokinin levels (Hussey, 1986).
4. It is promoted at low light irradiance and high temperatures.
5. It can be promoted by too intensive sterilization.
6. Some agars are more susceptible to vitrification, and these products should be avoided.
7. Vitrification may be lessened in some cases by changing the mixture of macro-salts. Changing from MS-medium to Lepoivre medium solves the problems of vitrification with *Malus* and *Prunus*.
8. Young soft plant material is more susceptible to vitrification.
9. The use of 2-phase media sometimes lessens or completely prevents vitrification.

In general it can be concluded from the research of Debergh (1983) and Ziv and Halevy (1983) that the vitrification is mainly a result of too high a humidity in the growth-tube, too low an agar concentration in the solid media, or to growth on a liquid medium. Debergh (1983) paid a lot of attention to the phenomenon of vitrification and concluded that through raising the agar concentration (to prevent this condition):

1. Vitrification was lessened as a result of reduced water uptake.
2. The axillary shoot multiplication was hindered.
3. The matrix potential changed (water uptake was more difficult).
4. The explant had less efficient contact with the medium (giving rise to the difficulty in water uptake).
5. The availability of macro- and micro-elements (added to the medium) change (dependant on the agar concentration).

6.4.3. Sugar

Sugar is a very important component in any nutrient medium and it's addition is essential for in vitro growth and development, because photosynthesis is insufficient, due to the growth taking place in conditions unsuitable for photosynthesis or without photosynthesis (in darkness). Green tissues are not sufficiently autotrophic in vitro. The CO_2 concentration in the test cube can also be limiting for photosynthesis (Gibson, 1967) and in practice it is very difficult and expensive to feed CO_2. In reality, as a result of poor gaseous exchange the CO_2 concentration in the tubes or containers may become too high and therefore toxic.

A concentration of 1–5% saccharose (a disaccharide) is usually used in vitro, since this sugar is also synthesized and transported naturally by the plant. Glucose and fructose may also be used. The sugar concentration chosen is very dependant on the type and age of growth material; for example very young embryos require a relatively high sugar concentration. Generally the growth and development increases with increased sugar concentration, until an optimum is reached, and then decreases at (very) high concentrations (Fig. 6.5). The growth of whole plants such as orchid seedlings is also greatly influenced by the sugar concentration (Fig. 6.6).

Saccharose (sucrose) that is bought from the supermarket is usually adequate. This is purified and according to the manufacturers analysis consists of 99.94% saccharose, 0.02% water and 0.04% other material (inorganic elements and also raffinose, fructose, and glucose). There is no indication that these other constituents can cause toxicity in vitro.

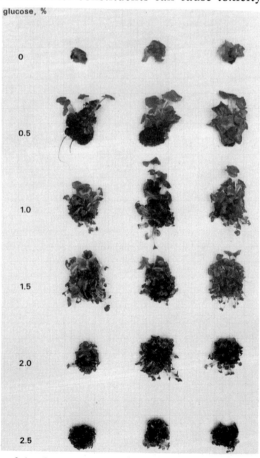

glucose, %

0
0.5
1.0
1.5
2.0
2.5

Fig. 6.5. The effect of the glucose concentration on adventitious shoot formation in callus tissues of *Begonia venosa* (Pierik and Tetteroo, 1986). Photograph taken 12 weeks after isolation.

0 % sucrose

0.5 % sucrose

1.0 % sucrose

2.0 % sucrose

Fig. 6.6. The sucrose (saccharose) concentration has an enormous influence on the growth of *Phalaenopsis* hybr. seedlings in vitro (Pierik and van Nieuwkerk, unpublished). Photograph taken 14 weeks after isolation.

As seen in Section 5.3.2, sugars can undergo changes as a result of autoclaving, in which pH is an important factor (Section 6.4.5). It must also be taken into account that during in vitro culture (e.g. with root culture) sugars can change in the medium: hydrolysis of saccharose as a result of invertase (from cell walls) or through extracellular enzymes (Burström, 1957; Weston and Street, 1968). Alteration of the sugar composition of the medium can also take place with callus cultures (George and Sherrington, 1984). Here it also appears that the presence or absence of IAA in the medium is very important. Other sugars can also become hydrolysed by plant tissues. Maretzki et al. (1971) showed that the extracellular hydrolysis of starch is possible during the growth of cell suspensions of sugar cane.

It is very difficult to answer the question, whether autoclaving of sugars is advisable or not for in vitro culture. There is conflict in the literature over this point, in some cases autoclaving is advisable (Ball, 1953) and in other cases not (Stehsel and Caplin, 1969).

6.4.4. *Mineral nutrition*

After sugars, minerals are the next most important group of nutrient materials for in vitro growth. There is a large choice of combinations of macro- and micro-salt mixtures. The most important mixtures are given Table 1. Chaussat et al. (1986) developed a computer program to calculate the quantities of salts which are necessary to make a nutrient solution the main characteristics of which are known (ionic proportions, total ionic concentration, pH). Or inversily starting from the composition of the nutrient solution given as salt masses, the program displays the characteristics of that solution. Concentrated stock solutions are normally used in the preparation of nutrient media, although ready-made media are also available. Stock solutions should be stored at room temperature in darkness. Complications can arise when making stock solutions (precipitation), and it is advisable to follow the instructions given with the standard media. In recent years NaFeEDTA ($25–38$ mg l^{-1}) has been used as the source of iron.

The mixture of macro- and micro-salts chosen is strongly dependant on the experimental plant. The Murashige and Skoog (1962) medium is very popular, because most plants react to it favourably. However, it should be appreciated that this nutrient solution is not necessarily always optimal for growth and development, since the salt content is so high. For example *Gerbera* is salt sensitive in vitro (Pierik and Segers, 1973), as are *Rhododendron* and *Kalmia* (Pennell, 1985). To counteract salt sensitivity of some woody species, Lloyd and McCown (1980), developed the so-called WPM (woody plant medium).

When choosing the macro- and micro-salt mixture the following should be borne in mind:

1. The total concentration is sometimes important (Fig. 6.7). The second part of Table 1 where ion concentrations are given in mmol l^{-1} clearly shows that Knop (1884) is a salt 'poor' and Murashige and Skoog (1962) is a salt 'rich' medium.
2. Although nitrogen (N) is sometimes partially provided in organic form (Section 6.4.9) it is usual to supply N in the form of NH_4^+ and NO_3^- ions. The total N requirement is $12–60$ mmol l^{-1} in which the NH_4^+ content varies from $6–20$ and NO_3^- content from $6–40$ mmol l^{-1}. Most plants prefer NO_3^- to NH_4^+ although the opposite

Table 1. Macro-salt content of the most important media (in mg l^{-1})

Components	Knop (1884)	Knudson C (1946)	Heller (1953)	Nitsch (1972)	Gamborg et al. B5 (1976)	Schenk and Hildebrandt SH (1972)	Murashige and Skoog (1962)
KNO$_3$	250	—	—	950	2 500	2 500	1 900
NaNO$_3$	—	—	600	—	—	—	—
Ca(NO$_3$)$_2$.4H$_2$O	1 000	1 000	—	—	—	—	—
NH$_4$NO$_3$	—	—	—	720	—	—	1 650
(NH$_4$)$_2$SO$_4$	—	500	—	—	134	—	—
NH$_4$H$_2$PO$_4$	—	—	—	—	—	300	—
NaH$_2$PO$_4$.H$_2$O	—	—	125	—	150	—	—
KH$_2$PO$_4$	250	250	—	68	—	—	170
KCl	—	—	750	—	—	—	—
CaCl$_2$.2H$_2$O	—	—	75	220	150	200	440
MgSO$_4$.7H$_2$O	250	250	250	185	250	400	370

Ion content of the most important media (in mmol l^{-1})

	Knop (1884)	Knudson C (1946)	Heller (1953)	Nitsch (1972)	Gamborg et al. B5 (1976)	Schenk and Hildebrandt SH (1972)	Murashige and Skoog (1962)
N-total	10.942	16.036	7.058	27.385	26.756	27.336	60.017
NH$_4^+$	—	7.567	—	8.994	2.028	2.608	20.612
NO$_3^-$	10.942	8.469	7.058	18.391	24.728	24.728	39.405
H$_2$PO$_4^-$	1.837	1.837	0.906	0.500	1.087	2.608	1.249
K$^+$	4.310	1.837	10.060	9.897	25.815	24.728	20.042
Ca^{2+}	4.234	4.234	0.510	1.496	1.163	1.360	2.993
Mg^{2+}	1.014	1.014	1.014	0.751	1.014	1.623	1.501
SO$_4^{2-}$	1.014	4.798	1.014	0.751	2.028	1.623	1.501
Na$^+$	—	—	7.964	—	—	—	—
Cl$^-$	—	—	11.080	2.991	2.325	2.721	5.986
Total	23.351	29.756	39.606	43.771	60.188	61.999	93.289

Micro-salt content of the most important media (in mg l^{-1})

	Knop (1884)	Knudson C (1946)	Heller (1953)	Nitsch (1972)	Gamborg et al. B5 (1976)	Schenk and Hildebrandt SH (1972)	Murashige and Skoog (1962)
FeCl$_3$.6H$_2$O	—	—	1.0	—	—	—	—
FeSO$_4$.7H$_2$O	—	25	—	27.8	27.8	15	27.8
MnSO$_4$.4H$_2$O	—	7.5	0.1	25	13.2	13.2	22.3
ZnSO$_4$.7H$_2$O	—	—	1.0	10	2.0	1.0	8.6
H$_3$BO$_3$	—	—	1.0	10	3.0	5.0	6.2
KI	—	—	0.01	—	0.75	1.0	0.83
CuSO$_4$.5H$_2$O	—	—	0.03	0.025	0.025	0.2	0.025
Na$_2$MoO$_4$.2H$_2$O	—	—	—	0.25	0.25	0.1	0.25
CoCl$_2$.6H$_2$O	—	—	—	—	0.025	0.1	0.025
NiCl$_2$.6H$_2$O	—	—	0.03	—	—	—	—
AlCl$_3$	—	—	0.03	—	—	—	—
Na$_2$EDTA	—	—	—	37.3	37.3	20	37.3

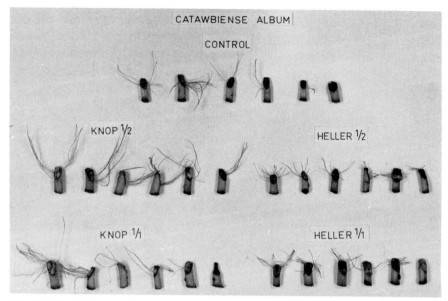

Fig. 6.7. The influence of two macro-salt compositions in (Knop, 1884; or Heller, 1953) at two concentrations (half and full strength) on adventitious root formation in stem explants of *Rhododendron* 'Catawbiense Album'. Control represents the absence of macro-salts. Optimal root formation occurs on the macro-salts of Knop (1884) at half strength. Photograph taken 8 weeks after isolation.

is true in some cases. It is necessary to find the right NO_3^+/NH_4^+ balance for optimal in vitro growth and development.

3. If the plant takes up NH_4^+ ions, the pH will decline and the agar may become liquid. As a result of the lower pH due to NH_4^+ uptake, the NH_4^+ uptake by the plant decreases and uptake of N in the form of NO_3^- takes over.

4. Table 1 showing concentrations of ions in $mmol\,l^{-1}$ shows that the K^+ requirement lies between 2 and $26\,mmol\,l^{-1}$ while those of $H_2PO_4^-$, Ca^{2+}, Mg^{2+} and SO_4^{2-} are small ($1-5\,mmol\,l^{-1}$).

6.4.5. *pH*

Little is known about the influence of the pH of a nutrient medium on in vitro growth. It is predicted that a pH in the range of 5.0–6.5 is suitable for growth with a maximum at about 6.0, since low pH's (lower than 4.5) and high pH's (higher than 7.0) generally stop growth and development in vitro.

If the pH is too low the following complications can arise (Butenko, 1968):

1. The auxin IAA and gibberellic acid become less stable.
2. The agar becomes too sloppy.
3. Particular salts (phosphate, iron salts) may precipitate.
4. Vitamin B_1 and pantothenic acid become less stable.
5. Uptake of ammonium ions is retarded.

The pH before and after autoclaving is different. If the starting pH is in the range 5.0–7.0 it generally lowers by 0.3–0.5 units (Skirvin et al., 1986).

Little can be found in the plant tissue culture literature concerning the use of buffers to control pH. Occasionally Sörenson phosphate buffer $(Na_2HPO_4 + KH_2PO_4)$ has been used successfully. However, the use of phosphate buffers can result in modification when added to the medium: Englis and Hanahan (1945) reported that glucose in a medium at pH 6.0 was partially converted to fructose as a result of autoclaving; also the conversion of saccharose to glucose and fructose during autoclaving appears to be pH dependent.

MES [2-(N-morpholino)ethane sulphonic acid] and TRIS[Tris(hydroxymethyl)methylamine] buffers (Bonga and Durzan, 1982) are also sometimes used in plant tissue culture. Many other buffers tested appear to be toxic, especially those capable of being assimilated by plants.

If the pH falls appreciably during plant tissue culture (the medium becomes liquid), then usually a fresh medium must be prepared at the desired pH. Often tissue cultures themselves buffer: a too high or too low pH being buffered to the desired value. It should be appreciated that a starting pH of 6.0 can often fall to 5.5 or even lower during growth (Skirvin et al., 1986).

If during medium preparation the pH is too low or too high then it may be corrected with diluted NaOH or HCl (0.1–1.0 M) respectively.

6.4.6. Osmotic potential

The osmotic potential of a nutrient medium is the sum of the osmotic potentials of the agar and other constituents (minerals, sugars etc.). Calculating the osmotic potential of a medium is very complicated and the molecular masses and degree of dissociation of salts are both important. In practice, the final osmotic potential can only be determined by measuring it. Sugar undoubtedly has a relatively higher influence on osmotic potential compared to macro-salts. It should be borne in mind that with sugars, a disaccharide (e.g. saccharose) is converted (hydrolysed) by autoclaving to 2 monosaccharides which will change its osmotic potential.

The respective contributions that sugars and macro-salts make to the

66

osmotic potential is very different in different media (Yoshida et al., 1973). The osmotic potentials in bar (1 bar $= 10^5$ Pascal) for macro-salts and sugars in a few media are respectively as follows: White (0.43; 1.46), Hildebrandt (0.67; 1.46), Heller (0.96; 4.05), Murashige and Skoog (2.27; 2.20).

If the osmotic potential is greater than approximately 3.10^5 Pascal ($= 3$ bar) growth and organ formation are stopped (Pierik and Steegmans, 1975a) as a result of the cessation of water uptake. The osmotic potential of a medium can be raised relatively simply by the addition of mannitol; this being a physiologically inactive substance (Greenwood and Berlyn, 1973). The addition of 1 M mannitol to a medium should, ideally, result in an osmotic potential of -22.4 bar. Nowadays polyethylene glycol is also added to media to change the osmotic potential.

6.4.7. Regulators

6.4.7.1. *Introduction.* Hormones are by definition, organic compounds naturally synthesized in higher plants, which influence growth and development; they are usually active at a different site in the plant from that where they are produced and are only present and active in very small quantities. Apart from these natural compounds, synthetic compounds have been developed which correspond to the natural ones. Hormones and these synthetically produced compounds are collectively called regulators, and are primarily responsible for the distribution of the compounds which the plant biosynthesizes. They determine the relative growth of all organs in the plant.

In in vitro culture of higher plants regulators, especially auxins and cytokinins, are very significant. It can be said that in vitro culture is often impossible without regulators. Whether an auxin and/or a cytokinin has to be added to a nutrient medium to obtain cell extension and/or cell division is completely dependent on the type of explant and the plant species. For example, explants which themselves produce enough auxin do not need extra auxin for cell extension and/or division. There are also explants which produce sufficient cytokinins and also need no extra cytokinins to be added to the medium. The following divisions can be made with respect to growth of cells, tissues, organs etc.:

1. Cultures which need neither auxin nor cytokinin.
2. Cultures which only need auxin.
3. Cultures which only need cytokinin.
4. Cultures which need both auxin and cytokinin.

For previously unstudied species a decision must be made as to whether tissues or organs of the plant concerned need regulators added for in vitro growth and development. The absolute and relative amounts of the different regulators (auxins and/or cytokinins) needed must then be determined. Other regulators (gibberellins and ethylene) may also be needed.

Regulators are added with the help of stock solutions. For example a stock solution of 100 mg l^{-1} equals 10^{-4} g ml^{-1}; other lower concentrations being obtained by dilution (1 ml of this stock solution per litre gives a final concentration of 10^{-7} g ml^{-1}). If 5–100 mg of the regulator is required per liter of the nutrient medium then it is added directly in solid form, since this cannot be achieved using a stock solution.

There are sometimes problems when dissolving different regulators in water. It is recommended that IAA, IBA and NAA are obtained in the more soluble K-salt form. It is also possible to dissolve these three auxins as acids with the help of 0.1 M KOH or NaOH. Cytokinins are also usually dissolved in 0.1 M KOH or NaOH, whereas gibberellins are dissolved in water with the help of sonification.

Stock solutions of IAA and kinetin should be stored in the dark since they are unstable in the light, consequently these two regulators are broken down (by light) in nutrient media. As far as is known, IBA, NAA and 2,4–D (auxins), and BA (cytokinin) are far more stable in the light. Lengthy storage of regulators introduces additional problems, for instance an

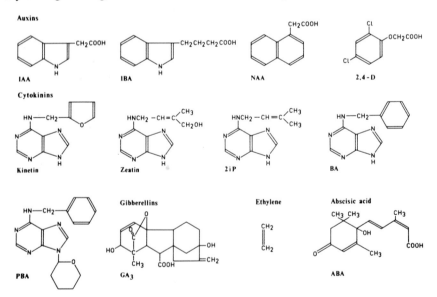

Fig. 6.8. Structural formulae of commonly used regulators: auxins, cytokinins, gibberellins (only GA$_3$ shown), ethylene and abscisic acid.

IAA in aqueous solution gradually becomes inactive; IAA is also easily broken down by enzymes (peroxidases and IAA oxidase).

The names and formulae of the regulators are given in Fig. 6.8, and in this Section only a short description is given of their most important effects.

If regulators, especially auxins and cytokinins are used habituation sometimes occurs. Habituation is the phenomenon in which in vitro cultures (which initially required a regular for growth and/or organ formation) after a period (after a few subcultures) no longer have a requirement for or require less regulators. This is often seen with callus cultures, and also in the case of formation of axiallary shoots under the influence of cytokinin (*Vriesea 'Poelmanii'*). Habituation is generally not a permanent change, in that if the plants formed from habituated tissues and explants are then isolated, these require regulators again. In other words; habituation changes are epigenetic in nature (changed gene activity at a different developmental stage of the plant). However, Horgan (1986) has put foward a theory that in some cases habituation is hereditable and not of an epigenetic nature.

In some cases (such as axillary shoot formulation in the case of the *Bromeliaceae*) it appears that isolating a shoot-tip from a mature plant after a few subcultures has a rejuvenating effect. This rejuvenation results in a considerable lowering of the regulator requirement, which however returns, as the culture becomes older. Therefore, the phenomenon of rejuvenation has some similarities to habitation.

6.4.7.2. *Auxins.* Auxins (IAA, IBA, NAA or 2,4-D) are often added to nutrient media. The naturally occurring auxin IAA is added in a concentration of $0.01-10 \text{ mg l}^{-1}$. The synthetic and relatively more active auxins (IBA, NAA, and 2,4-D) are used at concentrations of $0.001-10 \text{ mg l}^{-1}$. Figure 6.9 shows the influence of high concentrations of a weak auxin (IAA) compared to that of a low concentration of a strong auxin (NAA). Auxins generally cause: cell elongation and swelling of tissues, cell division (callus formation) and the formation of adventitious roots, the inhibition of adventitious and axillary shoot formation, and often embryogenesis in suspension cultures. With low auxin concentrations adventitious root formation predominates, whereas with high auxin concentrations root formation fails to occur and callus formation takes place. Use of 2,4-D should be limited as much as possible since it can induce mutations. At the same time 2,4-D can inhibit photosynthesis which does not occur with the auxins NAA, IBA and IAA.

Sometimes, the addition of auxins results in the promotion of seedling growth. Pierik et al. (1984) showed that NAA promoted root formation

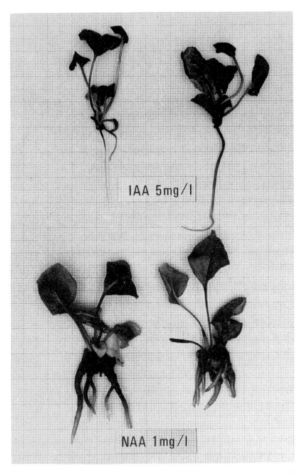

IAA 5mg/l

NAA 1mg/l

Fig. 6.9. The effect of a 'weak' auxin (IAA) in a high concentration and of a 'strong' auxin (NAA) in a low concentration on adventitious root formation in *Gerbera jamesonii* shoot cuttings in vitro (Pierik and Sprenkels, 1984). Photograph taken 6 weeks after isolation.

in seedlings of a selection of the *Bromeliaceae*, which resulted in growth stimulation of the germinated plants (Fig. 6.10).

6.4.7.3. Cytokinins. Cytokinins are often used to stimulate growth and development; kinetin, BA, 2iP, and PBA being in common use. They usually promote cell division, especially if added together with an auxin (Fig. 6.11). In higher concentrations (1–10 mg l^{-1}) they can induce adventitious shoot formation, but root formation is generally inhibited. They promote axillary shoot formation by decreasing apical dominance (Fig. 6.12) and they retard ageing.

| 0 | 5.10⁻⁷ | 10⁻⁶ |

Fig. 6.10. The effect of the NAA concentration in $g \, ml^{-1}$ on the growth of seedlings of *Vriesea splendens* in vitro. NAA promotes root formation, which results in promotion of the seedlings growth (Pierik et al., 1984). Photograph taken 20 weeks after isolation.

6.4.7.4. Gibberellins. This group of compounds is not generally used in the in vitro culture of higher plants. They appear in most cases to be non-essential for in vitro culture. GA_3 is the most used, but it must be borne in mind that it is very heat sensitive: after autoclaving 90% of the biological activity is lost (van Bragt et al., 1971). In general, gibberellins induce elongation of internodes and the growth of meristems or buds in vitro. They may also break dormancy of isolated embryos or seeds. Gibberellins usually inhibit adventitious root formation (Fig. 6.13; Pierik and Steegmans, 1975a), as well as adventitious shoot formation.

6.4.7.5. Other regulators

Oligosaccharins
Quite recently it was discovered (Albersheim and Darvill, 1985; Albersheim et al., 1986) that oligosaccharins (structurally defined fragments of cell wall polysaccharides) are chemical messengers with specific regulato-

71

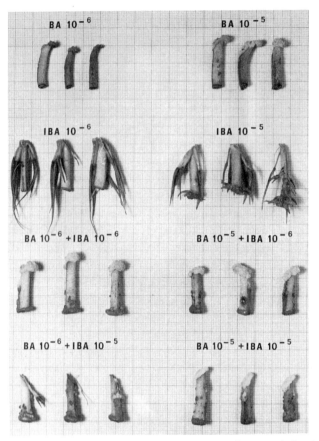

Fig. 6.11. The effect of BA and IBA (concentration in $g\,ml^{-1}$) on apolarly placed petiole explants of *Gerbera jamesonii* (Pierik and Segers, 1973). BA alone induces callus formation, whereas IBA alone induces adventitious root formation. When BA and IBA are both applied, callus formation is strengthened when the BA concentration is high (third row from above). Photograph taken 8 weeks after isolation.

ry properties. Oligosaccharins and released from the cell wall by enzymes; different oligosaccharins can regulate, not only by triggering the plant's defenses against pathogens and other types of stress, but also by regulating the rate of growth, and differentiation into roots, flowers and vegetative buds.

Abscisic acid
In most cases ABA has a negative influence (Pilet and Roland, 1971) on in vitro cultures (Fig. 6.14) and reports of promotion of callus growth and embryogenesis are probably incidental (Ammirato, 1983).

Fig. 6.12. The growth and development of excised shoots of *Vriesea* 'Poelmanii'. Above: control, no cytokinin in the medium. Below: with 0.08 mg l^{-1} BA in the medium. BA breaks apical dominance and as a consequence the axillary shoots develop abundantly. Photograph taken 8 weeks after isolation.

Ethylene

It is known that organ culture as well as callus culture are able to produce the gaseous hormone ethylene (Melé et al., 1982). Since tubes, flasks and especially plastic containers are sometimes entirely closed, care should be taken that there is no ethylene accumulation; since plastic containers can also produce ethylene, special care should be taken when using these. Flaming also forms ethylene (Hughes, 1981) and it is advisable not to do this in a laminar air-flow cabinet (where it makes little sense).

Melé et al. (1982) investigated the influence of different closure methods on the growth of shoot-tips of carnation and proposed that due to closure, ethylene accumulation takes place, resulting in growth inhibition. If KMnO$_4$ is added (in a separate tube) then 70% of the ethylene can be removed.

There are contrasting reports in the literature concerning the role played by ethylene in in vitro organogenesis (Huxter, 1981). Cases of no influence of ethylene on organogenesis and others where there is a posi-

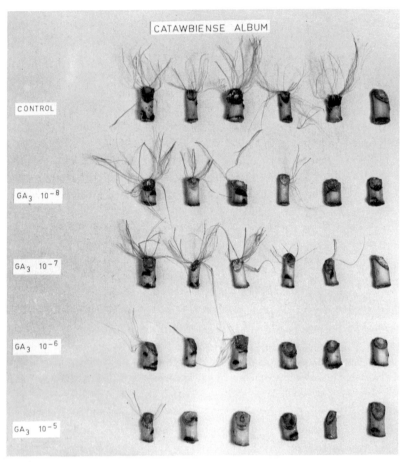

Fig. 6.13. Stem explants of *Rhododendron* 'Catawbiense Album' which have developed adventitious roots after treatment with IAA. When extra GA_3 is added, the formation of adventitious roots decreases with increasing concentrations of GA_3. In the bottom row with 10^{-5} g ml^{-1} GA_3 the formation of adventitious roots has been almost completely inhibited (Pierik and Steegmans, 1975a). Photograph taken 8 weeks after isolation.

tive effect have been reported. When tobacco calluses were grown in the light (Huxter et al., 1981) it was shown that ethylene production was greater when adventitious shoots were formed. They also showed that the ethylene formation depended on the time period after subculturing (more ethylene formation in the first 5 days than in the subsequent period of 6–10 days) and on the light/dark regime (more ethylene being formed in the dark than in the light). They proposed that ethylene lowered organi-genesis during the first 5 days of culturing, while the (visible) shoot differentiation was promoted in the subsequent 6–10 days after inoculation.

74

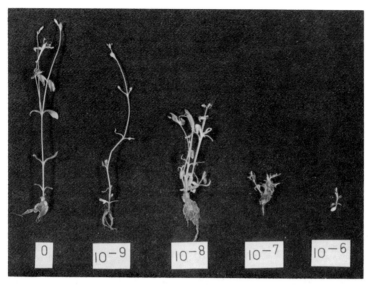

Fig. 6.14. Shoot tips of the long-day plant *Silene armeria*, grown in long days (16 h light per day). After stem elongation, flowering occurs in the test tubes. From left to right: increasing ABA concentrations slow down growth and stem elongation though at the highest ABA conc. (10^{-6} g ml^{-1}) a flower bud is still formed. The control (O) without ABA was the first to flower. Photograph taken 8 weeks after isolation.

Van Aartrijk (1984) and van Aartrijk et al. (1986) demonstrated with regenerating lily scale explants that ethylene and ethane production depend as much on the phase (in which the regeneration is taking place) as on the growth conditions: the number of adventitiously formed bulbs appears larger in proportion to the higher ethylene production during the first 2 weeks of culture. Ethylene (biosynthesis) therefore seems to have an important role to play in adventitious bulb formation. If ethylene biosynthesis in lily is blocked by AVG (amino-ethoxyvinyl glycine) the adventitious bulb formation is also stopped; addition of ethylene (1–10 ppm) during days 3–7 of the growth period promotes adventitious bulb formation, as does the addition of ACC (1-aminocyclopropane-1-carbonic acid).

Ethylene also appears to influence embryogenesis and organ formation in *Gymnospermae*. Verhagen et al. (1986) proposed that non-embryogenic callus of *Picea abies* produced 10 times as much ethylene as an embryonic callus. Kumar et al. (1986) showed that the build up of both ethylene and CO_2 during the first week of culture in Erlenmeyer flasks promotes adventitious shoot formation in excised cotyledons of *Pinus radiata*. When ethylene and CO_2 were eliminated from the flasks, shoot regeneration and growth were inhibited.

75

Sometimes the growth in vitro can be promoted by ethylene (Stoute-meyer and Britt, 1970; Palmer and Barker, 1973; Mackenzie and Street, 1970; Bouriquet, 1972). It appears that a certain ethylene concentration is necessary for induction of cell division, as was demonstrated by Mackenzie and Street (1970) for cell suspension cultures of *Acer*: 2,4-D induces ethylene formation.

6.4.8. *Vitamins*

One, or a few of the following vitamins are sometimes used in in vitro culture (alternative names and concentrations used in $mg \, l^{-1}$ in brackets): inositol (myo-inositol, meso-inositol; 100–200), vitamin B_1 (thiamine, aneurine; 0.1–5.0), Ca-panthothenate or panthothenic acid (0.5–2.5), folic acid (vitamin M; 0.1–0.5), riboflavin (lactoflavin, vitamin B_2; 0.1–10), ascorbic acid (vitamin C; 1–100), nicotinic acid (vitamin PP, niacin; 0.1–5), pyridoxin (adermine, vitamin B_6; 0.1–1.0), biotin (vitamin H; 0.01–1.00, para-aminobenzoic acid (0.5–1.0), tocopherol (vitamin E; 1–50).

The high concentrations of ascorbic acid which are sometimes added do not mean that the plant has such a large requirement. Vitamin C is used in such high concentrations as an anti-oxidant.

Most plants are able to synthesize vitamins in vitro, and perhaps it should be asked whether the frequent addition of vitamin mixtures during in vitro culture is always really necessary.

6.4.9. *Miscellaneous*

Polyamines
It has been shown that polyamines are involved in cellular differentiation and development during embryogenesis. Endogeneous levels of polyamines particularly putrescine and spermidine, were found to increase substantially during embryo formation in carrot. Embryogenesis was inhibited by the addition of polyamine synthesis inhibitors, and was restored with putrescine, spermidine or spermine. Similar effects were also described in *Medicago sativa* (Schneider and McKenzie, 1986). Rugini and Wang (1986) found that polyamines are cofactors in adventitious root formation. Another interesting point is that putrescine is capable of synchronizing the embryogenic process of *Daucus carota* (Bradley et al., 1985). It was concluded that polyamines and their associated enzymes are important in the control of plant growth and development. L-arginine appears to be a precursor of putrescine, and polyamines can block the conversion of methionine to ethylene. This suggests a control system in

which ethylene, arginine, and polyamines play an interrelated role in the control of somatic embryogenesis.

Phenylurea and it's derivatives

Since it was discovered that DPU (N,N'-diphenylurea) had cytokinin activity, many disubstituted ureas have been found that also exert cytokinin activity in various bioassay systems (Mok et al., 1986). In contrast to this Horgan (1986) proposed that DPU inhibited cytokinin oxidase, giving the appearance of increased cytokinin activity. In any case, it is certain that the above compounds possess their own definite properties which call for further investigation.

The most active compounds appear to be the pyridyl ureas and thidiazol ureas (thidiazuron and derivatives), which are up to 10,000 times more active than DPU, and more active than the naturally occurring adenine-type cytokinins such as zeatin. Thidiazuron has been reported to stimulate shoot proliferation in several woody species, whereas N-2-chloro-4-pyridyl-N-phenylurea (CPPU) caused a dramatic increase in the number of shoots in hardy deciduous azaleas (Read et al., 1986). Meyer and Kernsh (1986) also found that thidiazuron caused a marked increase in shoot proliferation of *Celtis occidentalis*. A concentration of 0.05–0.1 µM thidiazuron was more active than 4–10 µM BA.

A particularly interesting case is that of the poplar; Russell and McCown (1986) showed that thidiazuron (in a concentration less than 0.1 µM) is capable of inducing adventitious shoots in poplar calluses, which only show sporadic shoot formation using BA. However, interestingly axiallary shoot formation in poplar is severely inhibited in all thidiazuron treatments and it is necessary to transfer shoot-regenerating cultures to thidiazuron free media for shoot development.

Mixtures of compounds of vegetable origin.

Here we can classify such substances as coconut milk (the liquid endosperm of the coconut), orange juice, tomato juice, grape juice, pineapple juice, sap from birch, banana puree etc. In practice these mixtures should be avoided in research since:

1. The composition is almost or completely unknown.
2. The composition is very variable. For example, coconut milk (used at a dilution of 50–150 ml l^{-1}) not only differs between young and old coconuts but also between coconuts of the same age.

Work with coconut milk has produced some striking results. If used together with auxins it strongly induces cell division in tissues (Steward,

1958). Kovoor (1962) found that coconut milk contained a compound which was analogous to kinetin (a cytokinin). Letham (1974) proposed that mature coconuts contained 9-β-D-ribofuranosylzeatin (a cytokinin) and van Staden and Drewis (1975) showed the presence of both zeatin and zeatin riboside. In principle, cytokinin can be used in the place of coconut milk, and this is usually the case.

The preparation of coconut milk is as follows: a coconut (bought at the green grocers) is opened using an electric drill (drill three holes). The milk from each separate coconut is collected in a separate beaker and the liquid checked to ensure that it has not been 'decayed'. The milk from sound nuts is mixed together and filtered through cheesecloth. The filtrate is sterilized (15 min autoclaving at 120 °C), cooled and left to stand over night. The coconut milk is filtered off the following morning and stored in screw-top plastic containers in a deep freeze at -20 °C.

Fig. 6.15. The growth promoting effect of tryptone on in vitro grown seedlings of *Paphiopedilum callosum.* Above: no tryptone. Below: tryptone 1.5 mg l^{-1} (Pierik, unpublished results). Photograph taken 20 weeks after sowing.

Mixtures of compounds as the source of nitrogen and vitamins
The following compounds are used, as well as amino acids as organic sources of nitrogen: casein-hydrolysate ($0.1–1.0\,g\,l^{-1}$), peptone ($0.25–3.0\,g\,l^{-1}$), tryptone ($0.25–2.0\,g\,l^{-1}$), and malt extract ($0.5–1.0\,g\,l^{-1}$). These mixtures are very complex and contain vitamins as well as amino acids. Yeast extract ($0.25–2.0\,g\,l^{-1}$) is used because of the high quality of (B) vitamins. Figure 6.15 shows the promotion of growth of orchid seedlings by peptone and tryptone.

Amino acids as nitrogen source
The use of amino acids as an organic source of nitrogen is not usually required in modern media, where a proper balance between NO_3^-/NH_4^+ guarantees the nitrogen requirement. Previously amino acids (or mixtures of amino acids) were added as the NO_3^-/NH_4^+ was inadequate. L-glutamine is most commonly used as a nitrogen source, although adenine and asparagine may also be used.

Adenine(sulphate)
Adenine was used for the first time by Skoog and Tsui (1948) in the growth of stem explants of tobacco, where it stimulated the adventitious shoot formation. Nitsch et al. (1967) also added adenine to promote adventitious shoot formation. It is often added in so-called 'ready made' media, which are used in vegetative propagation (used at concentrations of $2–120\,mg\,l^{-1}$). It is possible to use adenine sulphate in place of adenine since it is more soluble in water.

Active charcoal
Active charcoal (used at concentrations of 0.2–3.0% w/v) is produced after wood is carbonized at high temperature in the presence of steam. It has a very fine network of pores with a very large inner surface area, on wich all sorts of substances (gases and solid compounds) can be adsorbed. Active charcoal (Merck nr. 2186 is often used) is purified to remove any remaining impurities. It is recommended that vegetable charcoal is used since this has a much higher percentage (95–99%) of active charcoal than charcoal obtained from an animal source.

The most important aspects of active charcoal are:

1. Adsorption of toxic brown/black pigments (phenol-like compounds and melanin) and other unknown colourless toxic compounds. Active charcoal is also able to adsorb toxic material in banana (Fig. 6.16).
2. Adsorption of other organic compounds (auxins, cytokinins, ethylene, vitamins, Fe and Zn chelates etc.) as given in a review article by

Fig. 6.16. Addition of banana homogenate without active charcoal (0.0%) has an inhibitory effect on the growth of *Phalaenopsis* hybr. seedlings in vitro. When active charcoal and banana homogenate are added the inhibitory effect of banana homogenate disappears; this is due to the fact that active charcoal adsorbs (an) inhibiting substance(s) present in the banana. High concentrations of active charcoal (0.6–1.0%) have an inhibitory effect (Pierik, van Nieuwkerk and Hendriks, unpublished). Photograph taken 14 weeks after sowing.

Misson et al. (1983). Johansson (1983) proposed that active charcoal adsorbed ABA.

3. The 'light climate' of the culture changes (darkening of the medium) as a result of which the root formation and growth can be modified (Klein and Bopp, 1971).

4. Active charcoal can promote somatic embryogenesis (Ammirato, 1983) and embryogenesis in anther cultures of *Anemone* and *Nicotiana* (Johansson, 1983). Evers (1984) and many other research workers have shown that the addition of active charcoal often has a pro-

motive effect on the growth and organogenesis of woody species. Beneficial effects of activated charcoal were also found in bulblet production of *Muscari armeniacum* (Peck and Cumming, 1986).
5. Activated charcoal stabilizes pH (Misson et al., 1983).
6. It is possible that active charcoal gives off substances which promote growth, but this has yet to be demonstrated.

Some plants have the unpleasant characteristic of exudating brown/black pigments upon wounding (usually oxidized polyphenol-like compounds and tannins) often making growth and development impossible. These can be prevented by (Anonymous, 1978a; Compton and Preece, 1986):

1. The addition of active charcoal to the medium (concentration of 0.2–3.0% w/v).
2. The addition of PVP to the medium (concentration of 250–1000 mg l^{-1}). PVP is a polymer which adsorbs phenol-like substances (Johansson, 1983).
3. Addition of so-called anti-oxidants such as citric acid, ascorbic acid, thiourea or L-cystine. These compounds prevent the oxidation of phenols.
4. The addition of diethyl-dithiocarbonate (DIECA) in the rinses, after sterilization, at a concentration of 2 g l^{-1}, and it's addition as droplets at the time of micro-grafting can also block oxidation phenomena (Jonard, 1986).
5. The addition of three amino acids (glutamine, arginine, and asparagine).
6. Frequent subculturing onto a fresh medium sometimes slowly stops the formation of pigments.
7. The use of liquid media in which it is easier and quicker to dilute toxic products.
8. Browning caused by photo-activation at the base of shoots can sometimes be eliminated by keeping the shoot bases in darkness during culture; penetration of light can be prevented by painting the outer sides of the jars or test tubes black up to the level of the medium, by wrapping the basis of the tubes with aluminium foil, or by applying a thin layer of inactivated charcoal and agar on the surface of the medium (Rugini et al., 1986).
9. A reduction of wounded tissue can result in a decrease in exudation.
10. A reduction of the salt concentration in the culture medium can reduce exudation.

11. Regulators play an important role in the darkening of the medium by oxidizing phenols. Omitting regulators can reduce exudation.
12. The soaking of explants in water before placing in culture has been an effective method of reducing exudation.

Phloroglucinol
Sometimes the phenolic compound phloroglucinol is added which inhibits the enzyme IAA-oxidase (responsible for the breakdown of IAA). Jones (1979), Hunter (1979) and Jones and Hopgood (1979) reported that phloroglucinol can promote axillary shoot formation and that applied together with auxin can work synergistically to promote adventitious root formation. The positive effects of phloroglucinol are not absolutely certain, since in later research it has been proposed that it can have no or even a negative effect on axillary shoot and adventitious root formation.

6.5. Commercially prepared media

Ready made media have been available since 1975 which has made the preparation of nutrient media in laboratories, schools, industry and in the home far easier. The producers are:

KC Biological Inc.,
U.S.A.

Flow Laboratories,
Scotland

Gibco,
U.S.A.

6.6. Storage of nutrient media

Nutrient media which must be stored for a long time should ideally be stored at low temperature (5 °C) in the dark. During prolonged storage care should be taken to prevent too much evaporation of water from the medium; therefore it is recommended that autoclaved media are packed in pre-sterilized and very clean plastic bags. Prolonged storage of media should be avoided if possible, since some substances are unstable in water (e.g. IAA).

7. Closure of test tubes and flasks

Test tubes and flasks need, on the one hand to be closed to prevent drying out and infection, while on the other hand a change of gases with the 'outside air' must be possible to avoid a shortage of O_2 and to prevent an accumulation of gases produced, such as CO_2 and ethylene. Each closure is a compromise between these two requirements. Closure can be obtained in the following ways:

1. Cotton wool plugs (Fig. 7.1). These plugs, previously used a great deal, are made by hand, which is very labour intensive. There are machines now available which make these to fit particular flasks and tubes. Previously, the cotton wool plugs were burned after inoculation, but this is no longer possible since the synthetic 'wool' gives off gases when burned which can be very harmful (Fig. 7.2).
2. Steristops (Fig. 7.1). These are porous cellulose stoppers, which can be pushed into a tube, like a cork.
3. Aluminium caps (Fig. 7.1), which are fixed on glass with a clip.
4. Transparent glass or synthetic tops, which can be autoclaved.
5. Tubes or flasks with screwtops. Care should be taken that they are not too tightly screwed, because then no exchange of gases can take place between the tube or flask, and the outside air.
6. Aluminium foil. This is used especially on large Erlenmeyer flasks.
7. Plastic (foam) plugs.

Seven methods of sealing have been discussed above, and (except 5 and 6) it is really necessary to have another second sealing procedure. This is necessary to prevent excessive evaporation and also infections (extra security). Vitafilm (Goodyear) and aluminium foil are often used as an extra seal in Western Europe, although other suitable films are available: Parafilm, PVC plastic film and polypropylene film. In principle the film to be used should allow the exchange of CO_2, O_2, ethylene etc., but not water vapour.

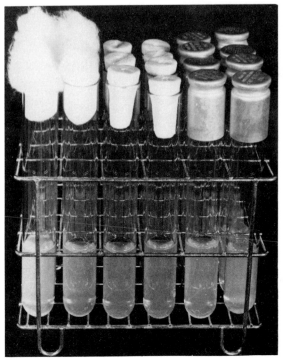

Fig. 7.1. Test tubes closed with (from left to right): cotton plugs, steristops and aluminium caps.

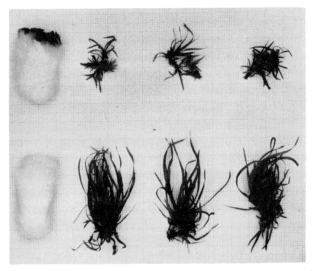

Fig. 7.2. Growth of seedlings of *Vriesea splendens* in vitro. Above: cotton plugs burned. Below: cotton plugs not burned (Pierik, unpublished results). Photograph taken 28 weeks after isolation.

The closure of test tubes or flasks can have far reaching consequences, which are laid out below:

1. If the closure is too tight then the gaseous exchange of CO_2, O_2, and ethylene etc. is barely possible or indeed completely impossible. The accumulation of CO_2 and/or ethylene can be detrimental. Aluminium foil and polypropylene both have the disadvantage that they close containers almost completely. Kimble manufactures polypropylene caps which act as a diffusible membrane with gases (with the exception of water vapour).
2. Synthetic tubes, boxes and caps can give off ethylene, which can be harmful when the container is too tightly closed.
3. If a tube is almost (completely) closed, the evaporation is greatly reduced, but the humidity in the tube is very high. Especially plants which are prone to vitrification (Section 6.4.2) need a lower humidity.
4. The choice of closure procedures also has an influence on the available light for the culture. Glass and synthetic caps allow light through from above in contrast to cotton plugs and aluminium foil.

Recently the water loss during culture in test tubes (closed by different methods) in continuous fluorescent light ($10 \ W \ m^{-2}$) at $25 \ °C$ was measured over 3.5 months (Pierik, unpublished results). The following water losses are given in $mg \ cm^{-2}$ test tube surface area:

Cotton plug + aluminium foil + Vitafilm	124
Cotton plug + aluminium foil	253
Cotton plug + Vitafilm	852
Metal cap + Vitafilm	710
Metal cap	930
Vitafilm	1105

Recent research has conclusively demonstrated that complete closure of tubes etc. has serious effects. De Proft et al. (1986) demonstrated that hermetically sealed Schott-flasks containing explants of *Magnolia soulangiana* had high CO_2 and ethylene concentrations, which reduced growth of the plant material and resulted in chlorotic plants. Woltering (1986), culturing *Gerbera* plantlets in polypropylene units, also concluded that in a closed (sealed) system without an ethylene absorbent (Ethy-sorb) the concentration of both ethylene and CO_2 was elevated, resulting in conditions detrimental to growth.

8. Care of plant material

Two sorts of plant material can be used for in vitro isolation: plants grown under controlled conditions in a greenhouse (or growth room), and those grown outside (grown in the open field). In practice, if an explant is isolated from a plant that has been grown outside, there is a much greater chance of infection taking place. Exceptions to this are tissues which are isolated from inside the plant: e.g. from the cambial zone of a shrub or tree, or from storage organs such as bulbs, tubers, rootstocks, etc. (Gautheret, 1959). If, despite the disadvantages material grown outside has to be used, then the following should be taken into account:

1. Buds should be used with bud scales, which are no longer dormant, although they have not yet burst.
2. Branches can also be used which have been held back (stored), and then forced in water inside.
3. Branches which have been grown outside can be enclosed in plastic, and only those parts which are subsequently formed used for isolation.
4. The youngest shoots formed outside should be used.

The very high percentage infection when using material grown outside, often makes it necessary to use material originating from a greenhouse or growth room. If outside material must be used it is advisable, where possible, to grow the plant inside in containers, to allow the formation of a new growth from which the explants for in vitro growth can be isolated.

Finally, to limit the infections when using material from a greenhouse or a growth room, the following guidelines should be followed:

1. Prevent insect infestations (aphids, red spider mite, whitefly, etc.), since these often carry diseases.

2. Prevent fungi and bacteria, where possible using (systemic) fungicides or bactericides.
3. Never water the plants from above, but only in the pot. Water is often an important requirement for the reproduction and spread of micro-organisms. This is particularly important for rosette plants.
4. Keep the humidity in the greenhouse as low as possible. Fungal and bacterial infections are more likely at high humidity.
5. Allow the plants to dry out before beginning sterilization and isolation.

The differences in contamination between plants which have been grown under different conditions has been well illustrated by Bourne (1977). He determined the average number of micro-organisms per tomato flower, and came to the following conclusions:

Source of flower	Non-disinfected	Disinfected	Contamination in culture after disinfection (%)
Field	1,300,000	92,000	100
Glasshouse	85,520	1,600	60
Phytotron	90	43	30

Especially if the plant material being used comes from trees of shrubs, which is only available for a short time for research purposes (often just before the buds burst), then it is important to use to correct procedures (use of plastic to prevent drying out during low temperature storage). Information on this can be found in Bonga and Durzan (1982).

In principle, only strong plants should be chosen for experimentation, since weak plants are more susceptible to disease. It is important that healthy plant material is used (free from viruses, bacteria, fungi, mites, etc.). For large scale production, care should be taken in choosing a suitable genotype; criteria such as rate of growth, earlines, flower production, flower colour, flower form, keeping qualities, etc. should be taken into account.

For these reasons experimental research is often carried out using extremely homogeneous plant material (the same developmental stage), which it is recommended is grown in a greenhouse or growth room, where it is easier to control growth and development. The influence of the starting material on in vitro growth and development is dealt with in more detail in Chapter 13.

9. Sterilization of plant material

9.1. Introduction

In principle there are 4 sources of infections: the plant (internal as well external), the nutrient medium (insufficiently sterilized), the air, and the research worker (inaccurate work). The most important of these is the plant itself, and plant material should be well sterilized before being isolated in vitro.

Before beginning sterilization, any remaining soil or dead parts etc. should be removed from the plants (or parts of the plants). This should be followed by washing in water if the external contamination is very bad (e.g. potatoes or rhizomes grown in the soil). Peeling results in the loss of the outer most layer (loss of the potato skin or the dry dirty layers of a bulb). After these steps, sterilization is begun, usually as follows: the organ is dipped into 70% alcohol for a few second (96% alcohol is too strong, resulting in excessive dehydration) to eliminate air bubbles, then sterilized for a 10–30 minutes in 1% NaClO containing a few drops of Tween 20 or 80, and rinsed (to remove hypochlorite) in sterile tap water (usually 3 times, for 2, 5, and 15 min respectively). After these steps work can begin on cutting the plant pieces, in sterile conditions (the laminar air-flow cabinet), and using sterile instruments (placed in 96% alcohol and then flamed).

If, despite 'good' chemical sterilization of the plant material, infections later occur, the cause of these are probably:

1. So-called internal infections, which are dealt with in Section 9.4.
2. Imprecise work (unwashed hands; table-top not sterilized with 96% alcohol; non-sterile forceps or scalpels; not sufficiently sterilized Petri dishes, paper and/or nutrient media; dirty laboratory coats, etc.). The use of face masks, covering the hair, and using sterile gloves can all contribute to a lower number of infections.

3. A defective laminar air-flow cabinet. This cabinet should be tested every year by the factory, with a so-called particle counter. The front filters should be periodically renewed.
4. The alcohol in which the instruments are placed prior to flaming is contaminated. It is advisable to renew this alcohol regularly.
5. The tubes, etc. containing nutrient media are not sterile on the outside. After sterilizing nutrient media, they should be stored under sterile conditions.
6. The floors are not regularly washed and disinfected. When entering the inoculation room plastic covers should be placed over the shoes, or the soles should be dipped in sterilizing fluid.
7. The room containing the laminar air-flow cabinets is not sterile. This can be avoided by blowing in sterile air, and by irradiating with UV light during the night.
8. Too many non-essential visitors are allowed into the inoculation room, infecting the floor and the air.
9. Infections often occur when isolating shoot-tips of rosette plants, because these are difficult to sterilize. To avoid this the shoots are often sterilized in stages: the shoot is washed, sterilized, a few leaves removed, sterilized again, etc. Generally speaking the chances of infection with shoot-tips is smaller if they are isolated when small. Preferably a meristem with only a few leaf primordia should be isolated; the disadvantage of this is that the chances of growth in vitro are smaller.
10. If the growth rooms are not kept clean (the floors not regularly cleaned and disinfected), infections can result, in some cases due to mites which can carry, especially, fungal infections with them. Since these mites can easily pass through cotton plugs and plastic film, the seriousness of this form of infection should not be underestimated. The eradication of mites is best carried out, over a few weeks with Vapona insecticide strips (dichlorovos) from Shell (Bonga and Durzan, 1982).

9.2. Chemical sterilization

Chemical sterilization (eradication of micro-organisms with the aid of chemicals) can be realized by:

1. Alcohol (ethanol); 70% alcohol is used for plant material since 96% alcohol dehydrates too much. When sterilizing tables and instruments 96% alcohol must be used, since 70% alcohol leaves a layer of water

after evaporation. When sterilizing plants dipping them for a few seconds in alcohol is not sufficient to kill all micro-organisms and after this (mainly used to eradicate air; alcohol dissolves also the epicuticular layer) they are usually treated with hypochlorite. Fruits can be externally sterilized by dipping in alcohol 96%, and then flamed; this is applied in the case of orchid sowing.

2. Bleach or sodium hypochlorite. This is available in most supermarkets or grocers and often consists of 10% active constituents ($=$ Na-ClO). Usually 1% NaClO is used, this being a solution of 1 part bleach (with 10% active constituents) and 9 parts tap-water, although higher concentrations e.g. 1.5–2.0% NaClO may also be used. If plants are particularly sensitive to bleach then it is advisable to use Ca-hypochlorite for sterilization.

3. Ca-hypochlorite or $Ca(ClO)_2$. This comes in the form of a powder which is mixed well with tap water; allowed to settle out and the clarified liquid (often filtered) used for sterilization. It is used for 5–30 min (concentration, $35–100 \, g \, l^{-1}$). Ca-hypochlorite enters plant tissues more slowly than sodium-hypochlorite, and can only be stored for a limited time since it is deliquescent (takes up water).

4. Sublimate or mercuric chloride ($HgCl_2$) dissolved in tap-water. This is an exceedingly toxic substance for plants as well as animals and man, and is used in a concentration of 0.01–0.05% (w/v) for 2–12 min. Rinsing must be very thorough, and because of it's extreme toxicity it should not be washed away down the sink.

Chemical sterilization can be made more effective by:

1. Washing the plant material very intensively with very clean water before beginning sterilization: allowing the water to regularly drain off. Hughes (1981) showed that this pre-rinsing, drastically reduced infections in the *Gesneriaceae*.

2. Placing the plant material in 70% alcohol for a few seconds, before chemical sterilization which results in the removal of air bubbles enabling the sterilizing liquid to have better access to the plant material.

3. The addition of Tween 20 or 80 (a wetting agent) to the sterilizing fluid (concentration 0.08–0.12%). A wetting agent lowers the surface tension, allowing better surface contact.

4. Stirring during utilization of bleach (with a magnetic stirrer).

5. Carrying out the sterilization with bleach under vacuum: this results in the loss of air bubbles and the sterilization process is carried out more efficiently.

The choice of sterilization time and the concentration of the bleach can be made depending on the particular conditions. They strongly depend on whether the surface layers of the explant being sterilized are to remain (gentle sterilization) or are to be cut off before inoculation (vigorous sterilization). It is sometimes possible that sterilization is accomplished after only 5 min treatment with 1% NaClO, and in other cases 30 min might be necessary. Lengthy sterilization can result in detrimental effects on the explant, and the correct timing and concentration of bleach should be decided for each individual experimental material. To give an idea of some of the concentrations and times used a few examples are given below:

Anthurium andreanum leaves: 30 min 1% NaClO
Hyacinthus scale tissue: 15 min. 1% NaClO
Rhododendron stems: 20 min 1% NaClO
Gerbera petioles: 15 min 1% NaClO
Freesia flower buds: 20 min 1% NaClO
Strelitzia leaves: 45 min 1% NaClO
Tulipa seeds: 30 min 2% NaClO
Phaseolus stems: 10 min 1% NaClO
Shoot tips of *Nephrolepis*: 5 min 1% NaClO (Soede, 1979).

Sometimes the surface of a wound may be coated in paraffin to prevent the penetration by the sterilizing fluid (into a stem), and /or to prevent bleeding (e.g. in the *Euphorbiaceae*).

9.3. Apparently sterile cultures

It should be borne in mind that seeming sterile cultures are not in fact always sterile. If the centre of infection is in the inner tissues of the plant, then the infection often only becomes apparent when the site of infection is cut open (during subculture), and makes contact with the medium. Contamination often only becomes evident after a few subcultures have been made. Poor growth and/or chlorosis can be an indication of an internal infection, e.g. bacteria such as *Erwinia carotovora* (Knauss and Miller, 1978). Apparently sterile cultures can also arise due to the fact that growth may take place on a relatively poor medium, on which micro-organisms seldom if ever develop. Infections only become evident if inoculation takes place onto a richer medium, or if mutation of the micro-organism takes place, allowing it to develop on a poorer medium.

92

To be certain that a culture is sterile, a shoot tip can be cut longitudinally, and placed with the cut surface on a rich medium. This can be prepared by the addition of 2–3% tryptone or peptone (a mixture of amino acids and vitamins), and within a few days there is usually an explosion of microbial growth (Knauss, 1976). This technique is not always sufficient since there is no good detection medium for many of the endogenous bacteria.

Internal contamination (see Section 9.4) in plants is often in the form of rod bacteria (particularly *Bacillus licheniformis* and/or *Bacillus subtilis*), which have been given the name of white ghost in the U.S.A. These saphrophytic bacteria are also often encountered in microbiological laboratories. They are well known because their spores are able to tolerate unfavourable periods (heat, drought, cold, UV-radiation, presence of sterilizing fluids). Sterilization of the air, floor and the plant material itself may be needed to rid the plants and the laboratory of the 'white ghost'.

9.4. Internal infections

Internal infections, which can be a considerable problem, are caused by micro-organisms present inside the plant itself, and cannot be eliminated by external sterilization. In principle there are two ways of combating this problem: meristem culture (since most of the micro-organisms are not present in the meristem), or by the addition of antibiotics to the nutrient media. Since meristem culture is very complicated and the addition of antibiotics to the media is largely ineffective, easily the best solution is the use of internally sterile plants!

The addition of antibiotics often leads to phytotoxic phenomena: such high concentrations of antibiotics are necessary that the growth and development of the higher plant is also inhibited. The use of antibiotics can also lead to the selection of a resistant micro-organism.

When used most antibiotics are added to the medium by filter sterilization. Stichel (1959) and Montant (1957) described the use of penicillin and achromycin, and other workers have used tetracyclin and 8-hydroxyquinoline. Staritsky et al. (1983) compared the efffects of different antibiotics (oxytetracycline, streptomycin, chloromycetin, penicillin-G, rifampicin, and gentamycin) on the culture of *Cryptocoryne* and *Cinchona* which were internally infected with bacteria. They concluded that only in the case of rifampicin were the bacteria inhibited, and at the same time there was no effect of the antibiotic on the growth and development of the explants. Young et al. (1984) found that no single antibiotic was effective against bacterial contaminants in shoot cultures of woody plants, and a combination of different antibiotics was more effective than any administered singly. More information concerning the use of antibiotics in overcoming internal infections can be found in George and Sherrington (1984). These authors reported that in some cases antibiotics are capable of promoting (sometimes dramatically) the growth rate of the cultured tissues.

Generally speaking the results of adding antibiotics to in vitro cultures of higher plants is not encouraging. Debergh (personal communication), has concluded that inhibiting the growth of micro-organisms by the addition of antibiotics is certainly not practical in the case of the commercial in vitro propagation of plants. The most effective method of overcoming internal infections is still meristem culture, as described by Tramier (1965) for gladioli.

9.5. Symbiotic cultures

Although in vitro culture of higher plants is by definition carried out under sterile conditions, the use can be made of micro-organisms in in vitro culture. If the germination and further development of orchid seeds is not possible asymbiotically, then so-called symbiotic germination can be brought about by the addition of the fungus with which the orchid lives symbiotically under natural conditions. Symbiotic culture may also be used in vitro to study the symbiosis between members of the *Papilionaceae* and nitrogen fixing bacteria, such as *Rhizobium* under controlled conditions.

10. Isolation, inoculation and subculturing

10.1. Introduction

It is obvious that inoculation and subculturing should also be carried out under the same sterile conditions as isolation. In a professional laboratory these take place in the laminar air-flow cabinet, but if there is not enough laminar air-flow space available (e.g. during practical courses), cutting etc., can be carried out between sterile filter paper.

It is advisable to wear a clean laboratory coat during preparations, and to wash your hands as well as washing the table top with 96% alcohol. Instruments such as scalpels, forceps, inoculation needles, etc. must be previously sterilized by immersing in 96% alcohol followed by flaming. Alternatively they can be put in a beaker containing glass balls at a temperature of 250 °C. This method (factory: S. Keller, Lyssachstrasse 83, CH 3400, Burgdorf, Switserland), has the great advantage that there is no need for flaming which can be a fire hazard. Sterilization of instruments without the use of alcohol can also be accomplished with the so-called Bacti-Cinerator, which has been developed in the U.S.A. This apparatus is electrically heated to very high temperatures, and it is only necessary to put the instruments inside for 5 seconds for them to be sterilized. As with the heated glass balls, care should be taken that the instruments are allowed time to cool before use.

If the alcohol is used for sterilization of instruments then it should be regularly changed since bacteria are sometimes able to survive emersion in alcohol, and the alcohol becomes contaminated with pieces of plant material and agar, etc.

Explants, can in principle, be cut in two different ways (Fig. 10.1): on a glass plate sterilized with 96% alcohol (this has the disadvantage that the knives quickly become blunt), or on (between) sterile filter paper (the knives stay sharp longer). It is very easy to work with two stacks of sterile papers: one for use in cutting the explants (regularly replaced) and one

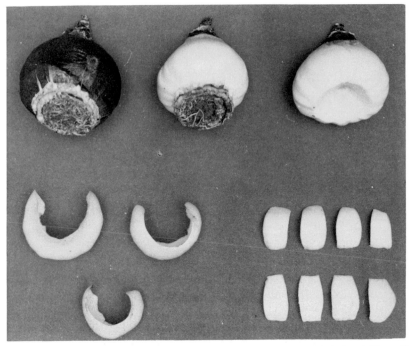

Fig. 10.1. Preparation of bulb scale explants from a hyacinth bulb. Scales (below left) are first sterilized, before explants (below right) are cut in the laminar air-flow cabinet.

for lying the sterile instruments on or between. To work efficiently a number of forceps, scalpels, etc. should be available.

10.2. Isolation

After sterilization and rinsing, the explant is laid on the sterile filter paper or glass plate using sterile forceps. If the cut surfaces have been in contact with bleach, the effected parts are first removed using a sterile scalpel. Sterilized seeds (if no embryos need to be isolated) can be directly inoculated without any further treatment.

It if often necessary to cut out a standard amount or volume of tissue, and to make this more easy to realize, graph paper (coated in plastic) is available. Other necessary pieces of apparatus are the cork borer, the cutting apparatus developed by Bouriquet (1952), and scales for sterile weighing on aluminium foil. When cutting explants it should be borne in mind that the volume made available can have important consequences (see also Chapter 12): the amount of food reserve, the cut surface area (ethylene production).

The preparation of meristems under a binocular microscope is usually accomplished with the use of a piece of razor blade fixed on an inoculation needle holder. Pointed razor blade pieces can be made sticking cellotape on the blade, and then cutting pointed pieces with sharp scissors or breaking them off with pincers. The cellotape is then removed and the pointed pieces mounted on the inoculation needle holder.

10.3. Inoculation

During inoculation the test tube or flask containing the solid medium, should in principle, be held horizontally. This strongly reduces the number of infections, particularly when not working in a laminar air-flow cabinet. Flaming the neck of the test tube or flask (especially in a laminar

Fig. 10.2. Adventitious rooting of excised *Rhododendron* 'Catawbiense Album' stem explants is strongly promoted when the explants are placed upside down (apolarly) in darkness on a medium containing auxin and sugar. Adventitious roots are always formed at the basal sides of the explants (Pierik and Steegmans, 1975a). Photograph taken 8 weeks after isolation.

air-flow cabinet) should be avoided, since this can result in ethylene penetration into the test tube or flask (Hughes, 1981).

The method of inoculating on solid media strongly depends on the experimental material. Seeds are usually placed on rather than in the medium, which results in oxygen deficiency. This also applies to meristems, which are inoculated on the medium using an inoculation needle (dampened with sterile agar), or a piece of razor blade mounted on an inoculation needle holder. Explants (e.g. a piece of pith tissue) are usually pushed half-way into the agar. Care should be taken not to push shoot tips far into the agar (oxygen deficiency will result). Explants retain their polarity after inoculation: the physiological upper-side remaining the upper side, etc. It is very important when regenerating organs to know how the inoculation has been carried out: polar (straight up, with the basal side of the explant in the medium), or apolar (upside down, with

Fig. 10.3. Soon after isolation a shoot of *Pelargonium* hybr. has excreted a halo of black substances into the culture medium.

the basal side above the medium). Adventitious roots are mainly formed on the basal side of the explant (Fig. 10.2), which results in better adventitious root formation with apolar inoculation, as expected from the better oxygen availability. If the original plant material is orthotropic or plagiotropic then these usually remain so in vitro. For the formation of axillary shoots on isolated shoot tips, it is sometimes best to lie the shoots horizontally on the medium, which promotes the formation of side shoots (Frett and Smagula, 1983).

As has been mentioned earlier shortly after in vitro isolation a brown/black halo of pigment can exudate in the agar (Fig. 10.3). Ways of overcoming this problem can be found in Section 6.4.9.

10.4. Subculturing

Subculturing can be necessary for a number of reasons:

1. The nutrient medium is exhausted (deficiency phenomena).
2. The nutrient medium dries out (resulting in too high salt and sugar concentrations).
3. Growth has filled the tube or flask.
4. The material is needed for further propagation.
5. Brown and/or black colouring appears in the agar: plant tissues sometimes give off toxic substances during the first few weeks, which diffuses into the agar or liquid medium.
6. It is needed to give the isolated material a different growth and development pattern, on a known nutrient medium.
7. The medium has become liquid due to a lowering of the pH by the plant.

Subculturing is carried out as follows:

1. The tube or flask is externally sterilized with 96% alcohol (on a cotton wad).
2. Any aluminium foil or film and then the cotton wad (or steristop) are removed from the tube or flask in the laminar air-flow cabinet.
3. The explant or callus clump is taken out and put in a sterile Petri dish or on (between) sterile filter paper.
4. After any cutting out the material is inoculated onto a new nutrient medium. When cutting pieces out, strong homogeneous (not necrotic etc.) material is selected.

11. Mechanization

If machines are used for in vitro culture, then it is certain that liquid media are being used. This has the following consequences (Bonga and Durzan, 1982):

1. The disadvantages of agar are no longer encountered (natural product with a complex and variable chemical composition), which results in more uniform cultures.
2. Division, growth and vegetative propagation are more rapid on a liquid than on an agar medium. Shakers and fermentors are already in use in the vegetative propagation of lilies (Takayama and Misawa, 1982).
3. In comparison to agar media, cell division in liquid media is far easier to synchronize, making biochemical research much easier.

Rotating and shaking machines are generally used when growing cells, cell suspensions, tissues, protocorms, meristems and shoot tips in a liquid medium. If cells etc., are grown immersed in a liquid medium, it is usually vital to keep them moving by using such a machine. This promotes gaseous exchange (oxygen, carbon dioxyde, ethylene), eliminates the effects of gravity, and stops the formation of nutrient and hormonal gradients (Street, 1973), giving much stronger cell division, growth and/or propagation. The machines are usually installed in special rooms.

There are a variety of machines to choose from:

1. Slow or rapid. Examples of slow machines are the so-called orchid wheel (Fig. 11.2), and the Steward machine, while most rotary shakers are more rapid (Fig. 11.1).
2. Some machines allow for periodic immersion of the cells etc. (the Steward machine), while others keep the cells etc. continually in the liquid medium.

Fig. 11.1. Rotary-platform-shaker. This rotating machine has three shelves each with 30 Erlenmeyer flasks of 300 ml (containing 100 ml of liquid culture medium); metal clips are fitted on the shelves to hold the aluminium covered flasks which are also kept in place by a clamp.

3. Combinations of 1 and 2. The orchid wheel (Fig. 11.2), for example, is a slow machine which allows meristems and protocorms of orchids to remain continuously in the liquid medium.

Slow machines usually have a rotation speed of 2–4 r.p.m. The orchid wheel is at an angle of 45° and the Steward machine at 12–15°. Two sorts of glassware can be used with the Steward machine: large round bottomed flasks with nipples (Fig. 6.2) and tumblers. Test tubes (put in test tube baskets; Fig. 11.2), or 100 ml Erlenmeyer flasks (fixed with clamps or elastic; Fig. 11.3), are used with the orchid wheel. The orchid wheel is also used with small flasks when propagating shoots; in Fig. 11.3 a wheel

Fig. 11.2. Slowly rotating wheel used in the orchid industry to propagate orchids by means of meristem and protocorm culture. The test tubes (with liquid medium) are placed in test tube racks, which are fitted on the sloping turntable.

is depicted with Erlenmeyer flasks, in which for exemple the *Bromelia-ceae* are vegetatively propagated by axillary shoot formation.

The so-called rotary shakers generally rotate much more quickly (30–150 r.p.m.). This sort of machine (with variable speed), is often made up of three layers above each other (Fig. 11.1), with clamps or elastic mounts on each layer, in which the wide mouthed Erlenmeyer flasks are placed. The Erlenmeyer flasks (usually 300 ml flasks containing 100 ml of nutrient medium), are covered with aluminium foil. Rotary shakers are mainly used for the growth of cells, cell suspensions or calluses, and virtually never with meristems, shoot tips or protocorms. The speed of rotation used with this type of machine depends on the type of plant material being used, since too rapid rotation can cause damage to cells and cell aggretates.

Fig. 11.3. Slowly rotating turntables on which metal clips are fitted to hold Erlenmeyer flasks (which are also kept in place by a clamp) of 100 ml (with 30 ml liquid culture medium). This turntable is used to propagate *Bromeliaceae* and other plants.

There are many other types of machines apart from those described above. For biotechnological applications, cells of higher plants are grown in so-called fermentors, in which aeration with sterile air is realized. Styer (1985) has recently published a review on machines suitable for use with vegetative propagation of plants (especially by somatic embryogenesis). A description of the available types of machines together with their advantages and disadvantages is given in the handbook by Street (1973), which also illustrates some of the different types of growth curves of cells and cell suspensions observed.

Below are given some of the technical terms used when growth takes place in liquid media, with the help of a machine (from King and Street, 1973):

Batch culture
Cells are grown in an open system with a definite amount of nutrient medium available which is not renewed; the growth stops when one or more of the nutrient requirements is exhausted. In this culture system there is no balanced growth.

Continuous culture
Cells growing in an open system in a constant amount of nutrient

medium which is renewed by influx; the amount of incoming medium being exactly equal to the amount of efflux.

Open continuous culture
A continuous culture, in which the influx of fresh medium is equal to the efflux of old medium together with the cells. In this sort of culture cells are lost with the old medium. A 'steady state' is achieved when the loss of cells is equal to the new cells formed.

Closed continuous culture
This is a continuous culture where the amount of efflux of old medium and influx of fresh medium is equal, but there is no efflux of cells with the old medium.

Chemostat
An apparatus in which continuous culture is carried out, where the growth rate and cell density are kept constant by a fixed rate of influx of a nutrient limiting growth.

Turbidostat
An apparatus in which a continuous culture is carried out, where fresh medium is added when a particular cell density is reached. The cell density is pre-determined (biomass), and kept constant by the efflux of cells out of the system with the old medium.

Recently, Tisserat and Vandercook (1985), described a new automated plant culture system to grow tissues, organs, and whole plantlets under sterile conditions. This system incorporates independent or multiple concurrent growth of cultures. The automated plant culture system (APCS), consists of silicone tubing, 2 impeller pumps, 2 medium reservoir bottles of glass, a 3 way stainless steel valve, a plant culture chamber and an interface module containing relay bonds. Control of ACPS is through interfacing with a micro-computer, which controls medium introduction, evacuation, and replenishment in a sterile environment. The ACPS is relatively inexpensive to construct, and provides a labour-saving, long term method of culturing plant in vitro. In a subsequent article Tisserat and Vandercook (1986) report the application of their ACPS to long-term culture of the orchid *Potinara*.

12. The influence of plant material on growth and development

The experimental material itself, as well as the nutrient medium and physiological growth factors, can influence growth and development in vitro. The influence of the plant material can be summarized as follows:

1. *Genotype.* There is a wide range of regenerative capacity in the plant kingdom. Dicotyledons generally can regenerate better than monocotyledons, and gymnosperms have very limited regenerative capacity (except when juvenile). Amongst the dicotyledons, *Solanaceae, Begoniaceae, Crassulaceae, Gesneriaceae* and *Cruciferae* regenerate very easily.

   There are very great differences in cell division and regenerative capacity between plants within a single species (Fig. 12.1). If a species regenerates organs easily in vivo (*Saintpaulia ionantha, Begonia rex, Streptocarpus* hybrids), then almost the same can be expected in vitro. However, there is sometimes a sharp contrast between the regenerative capacity in vivo and in vitro: for instance, it is almost impossible to have adventitious shoot formation in vivo, from leaf cuttings of *Kalanchoë farinacea*, whereas this is perfectly possible in vitro, perhaps due to a better uptake of regulators.

2. *The age of the plant.* Embryonic tissues usually have high regenerative capacity, and for instance with cereals, embryos and seeds are often used as experimental material for tissue culture. As a plant becomes older its regenerative capacity often decreases, and parts of juvenile plants are preferred to those from adult especially in the case of trees and shrubs.

   A few striking examples of differences in cell division and regeneration between juvenile and adult plants in vitro are: *Hedera helix* (Stoutemeyer and Britt, 1965), *Lunaria annua* (Pierik, 1967), and *Anthurium andreanum* (Pierik et al., 1974a).

107

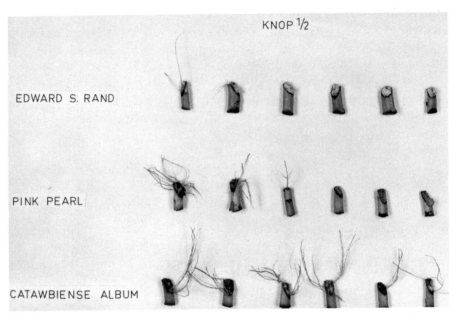

EDWARD S. RAND

PINK PEARL

CATAWBIENSE ALBUM

Fig. 12.1. Regeneration of adventitious roots from apolarly placed stem explants of three *Rhododendron* cultivars. Basic medium contained half strength Knop (1884) macro-elements. From top to bottom: cultivars with poor, moderate and fair formation of adventitious roots respectively (Pierik and Steegmans, 1975a). Photograph taken 8 weeks after isolation.

When isolating meristems and shoot-tips (shoot initials) it must be borne in mind that juvenile shoot-tips remain juvenile in vitro, while an adult shoot-tip remains adult. Sometimes through repeated subculturing of shoot-tips or more especially meristems, an adult meristem gradually (step-by-step) takes on juvenile characteristics. This rejuvenation results in increased cell division and regeneration, as shown by Hackett (1985), for *Pinus pinaster, Sequoia sempervirens, Vitis vinifera, Malus sylvestris, Rhododendron* hybrids, *Thuja occidentalis*, and *Cryptomeria japonica*. Rejuvenation can also be achieved by adventitious shoot formation, and this, together with other methods of rejuvenating tissues can be found in Section 20.4.1.

3. *The age of the tissue or organ.* Young, soft (non-woody) tissues are generally more amenable for culture than older woody tissues, although a large number of exceptions can be found in the literature.

When isolating pieces from a petiole it is found that a very young petiole often regenerates better than a young petiole (Fig. 12.2), and the young one better than an old one. As the organ from which the

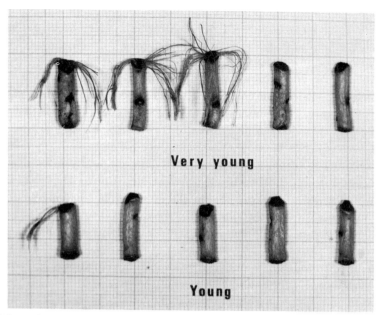

Very young

Young

Fig. 12.2. In vitro adventitious root formation on apolarly placed stem explants of *Rhodod-endron* 'van Weerden Poelman' is promoted by taking very young (soft) petioles. The young petioles are 5 days older than the very young petioles (Pierik and Steegmans, 1975a). Photograph taken 8 weeks after isolation.

explant is isolated becomes older, cell division and regeneration decrease.

It is especially noticeable that the regenerative capacity of different species (irrespective of age) increases during flowering: parts of young inflorescences are sometimes strongly regenerative as shown with *Freesia* (Bajaj and Pierik, 1974), *Lunaria annua* (Pierik, 1967), *Ranunculus sceleratus* (Konar and Nataraja, 1969), some *Amaryllidaceae* (Pierik et al., 1985; Pierik and Steegmans, 1986a), *Primula obconica* (Coumans et al., 1979) and *Rhododendron* (Meyer, 1983).

4. *The physiological state.* This has a strong effect on in vitro cell division and regeneration. In general, parts from vegetative plants regenerate more readily in vitro than parts from generative plants, although a few exceptions can be found in 3 above. Explants from scales of vegetative lilies regenerate better than those from generative lilies (Robb, 1957). Parts from juvenile plants regenerate more readily than parts from adult plants. Buds taken from plants (especially trees and shrubs), which are still in a resting state (late autumn or early winter), are more difficult to culture in vitro than buds from plants which are no longer dormant (taken in spring, just before they

109

are about to burst into growth). Dormancy is also an important consideration in seeds.

5. *The state of health.* If the plant is healthy at the time of isolation, then the in vitro culture is far more likely to be successful. If a choice must be made from cloned individuals then the healthiest should be chosen as experimental material, since this also has an effect on the percentage infection after isolation.

6. *Effect of different years.* If explants are isolated from plant material obtained from plants grown in the open field, then differences occur depending on the conditions encountered during the year, such as whether the winter was severe (facilitating the breaking of dormancy), the dryness of the summer (poor growth if insufficient water was available), or insufficient light during the growing season (less food reserves stored).

7. *Growth conditions.* If material is collected that has been grown under natural daylength and light conditions, then this will react differently to material that has been grown in a greenhouse. In general, material from a greenhouse (more elongated and etiolated) regenerates more readily than that from outside, a good example being *Rhododendron* (Pierik and Steegmans, 1975a). In the case of winter flowering begonias the growh conditions (daylength; temperature) have a definite effect on the formation of adventitious roots and sprouts.

8. *Position of the explant within the plant.* Topophysis is the phenomenon whereby the position of the explant in the plant influences the in vitro growth and development after isolation. For example, the higher the shoot is isolated from a tree, the lower the probability that adventitious roots will be formed; the pieces from higher up being more adult than those from lower down. An example of topophysis was described by Evers (1984). He found when working with *Pseudotsuga menziesii* that shoot initials isolated from positions low down on the tree showed better development in vitro, and that terminal buds grew faster than axillary buds.

Similarly, if explants are isolated and their original position in the plant noted, gradients of regeneration often become apparent. For instance it becomes obvious that a hyacinth forms adventitious bulbs more readily on an explant isolated from the base, rather than the top part of the bulb scale. The same is true for bulb scales of lily (Robb, 1957). Gradients of regeneration were also shown in tobacco (Aghion-Prat, 1965), *Lunaria annua* (Pierik, 1967), and *Brassica carinata* (Jaiswal, 1986). It is worth noting that calluses, arising from explants originating from very different parts of a plant, such as roots, shoots, petioles etc. can also react fully identical in vitro (Barker, 1969).

110

9. *Size of the explant.* Generally speaking it is far more difficult to induce growth in very small structures such as cells, clumps of cells, and meristems, than in larger structures such as leaf, stem or tuber explants. The isolated part of the plant has it's own supply of food reserves and hormones and it is obvious the larger the plant part the easier it is to induce growth and regeneration (Fig. 12.3). Plant parts containing a large amount of food reserves such as tubers, bulbs etc. generally regenerate more readily in vitro than those containing less reserves. When larger explants are isolated, then the addition of nutrients (sugar, minerals) and regulators can have less effect.

When cutting explants (small versus large), it should also be taken into account that the percentage of wounded to not wounded surface area can influence regeneration. The influence of wounding (resulting in ethylene formation) on the regeneration of lily bulb-scale explants has been clearly documented by van Aartrijk (1984).

By analogy to the above points 'mass-effects' play an important role in the growth of cells and orchid seeds. Care should be taken to have a particular cell density when beginning cell culture on e.g. agar plates or no cell division will take place. Orchid seeds germinate much better in vitro when they are sown thickly, and the subsequent growth and development is better if grown close together rather than

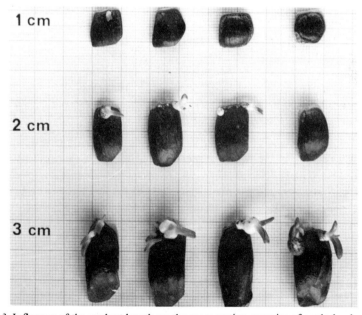

Fig. 12.3. Influence of the explant length on the regenerative capacity of apolarly placed bulb scale explants of hyacinth (Pierik and Ruibing, 1973). Photograph taken 12 weeks after isolation.

singly. This 'mass effect' is also known as the 'community effect' in the literature.

10. *Wounding.* When cutting explants it should be borne in mind that the area of wound tissue can be of great importance. Increasing the wound surface area increases the uptake of nutrients and regulators while at the same time increasing ethylene production. If there is an anatomical barrier (e.g. a layer of sclerenchyma vessels) in an explant for the formation of adventitious roots, then it is possible to break this by wounding (Fig. 12.4).

11. *Method of inoculation.* Explants can be placed on the nutrient media in different ways: polar (straight up, with the physiological base in the medium) or apolar (upside down, physiological base out of the medium and top in the medium). Roots and shoots generally regenerate more easily and more rapidly with apolar inoculation (Pierik and Steegmans, 1975a). Although this better regeneration may be a result of an improved oxygen supply, there may be other factors which play a role. Explants which are apolarly inoculated have sub-

Fig. 12.4. The effect of wounding on root regeneration of apolarly placed petiole explants of *Rhododendron* 'Pink Pearl'. In 1 the explants are cut transversely, in 2 cut obliquely, whereas in 3 a strip of bark is removed from the whole length of the explant. Root regeneration is promoted by the removal of the bark strip (Pierik and Steegmans, 1975a). Photograph taken 8 weeks after isolation.

stances accumulated at the basal end which cannot diffuse into the agar, since it is not in contact with the medium.

When as is the case for all the *Amaryllidaceae* (Pierik et al., 1985), regeneration of organs only takes place at the base of the bulb scales, then the method of inoculation is of particular importance; apolar inoculation results in better adventitious bulb formation than with polar inoculation.

12. *Nurse effects.* Nurse effects are often mentioned in the literature especially when discussing the growth of cells or cell aggregates on agar plates. This describes the effects when a few clumps of callus tissue are placed in the middle of a cell population. The callus gives off substances which diffuse into the medium and have a positive effect on the cell division of the individual plated cells.

13. *Preparation.* The physiological condition of the starting material can have an important influence on the in vitro growth of explants; we have seen many examples of different treatments in this Chapter. Another possibility is to artificially modify the physiological state of the starting material; this is possible via (Debergh, 1986):
   — Spraying the mother plants with regulators (e.g. cytokinins).
   — Injection of the starting material with regulators.
   — Putting explant sources in a forcing solution (with sugars, BA, GA_3, etc.).
   — Incubation of primary explants in solutions e.g. containing different BA concentrations.

PHILIPS

Artificial lighting for the irradiation of plants is an established fact in modern-day horticulture.

The grower has become independent of daylight and been able to increase the productivity of his enterprise in terms of both more and better plants and shorter cultivation times. In addition, he can cultivate plants to be available for the market at the most favourable time.

Through the years, Philips has played a major role in developing the numerous opportunities that exist in applying artificial lighting to plant growth, in whatever form.

Philips lighting in growth industry.

Horticulturalists are growing tissue cultures with the aid of 'TL' HF high-frequency fluorescent lighting, as illustrated above. Electronic ballasts enable the lighting level to be regulated and so provide ideal irradiance conditions for optimum growth.

In conventional greenhouses, SON-T high-pressure sodium lamps in SGR 200 luminaires offer an economic "after-hours" lighting solution for a wide variety of plant types. Here, rows of pot plants are being irradiated during the winter months.

Whatever the horticultural application, Philips can assist in putting forward suitable artificial lighting proposals.

Philips: Lighting Leadership Worldwide.

13. The influence of physical factors on growth and development

13.1 The culture room

It is necessary, both for research and for practical application of in vitro culture, that a growth room is available in which the light, temperature, etc. can be controlled. In most cases one growth room is insufficient, especially when working with different species having very different temperature requirements, or with one species under different temperature regimes. It should be well insulated, so as not to be effected by external temperature changes. Lighting is usually supplied by fluorescent tubes the chokes (control equipment) of which are (is) mounted outside the growth chamber, since if they are inside the chamber extra cooling is necessary to dissipate the heat given off. When in use the fluorescent tubes produce so much heat that the culture room needs to be cooled, the amount of cooling necessary depending on the number of fluorescent tubes in use. The temperature of a well insulated chamber will certainly remain at the desired temperature during the dark periods and if badly insulated heating may be necessary to maintain the desired temperature. The temperature control (cooling and heating), is achieved with the aid of a thermostat.

The temperature can be controlled by two different methods: a cooling and warming element in each growth chamber, or the pre-cooled or pre-warmed air is passed into the different growth chambers from a central regulation point. The second method of temperature control is used especially in large laboratories, with many growth chambers. A big advantage of this system is that the excess heat production from one chamber (during the light period) can be used to warm other chambers when necessary (during the dark period), but it is then a prerequisite that the light periods in the chambers are different. In some commercial laboratories a system is utilized, so that half of the growth chambers are completely out of phase (half at any one time having a light period, while the other half are

Fig. 13.1. Culture room, in which light and temperature can be regulated. The staging is made from Dexion and fluorescent light fittings are suspended between the shelves. The refrigeration unit is situated against the ceiling above the growth racks. The test tubes stand in holes that have been drilled in wooden blocks. For research purposes it is particularly useful to have a table available in the culture room for making observations.

in darkness). In this way the need for cooling and warming is reduced to a minimum. In such a system the control equipment for the lamps is usually mounted within the growth chamber.

Whatever system of temperature control is chosen (in or outside the growth chamber), it is necessary to make sure that in the chambers, and particularly between the growth racks there is good ventilation. If the different layers of the growth racks are more or less closed off from each other (Fig. 13.1), with a resulting lack in ventilation then the temperature may locally become too high due to the fluorescent tubes. This can have serious consequences for some species.

Fluorescent tubes can be installed in two different ways. They can be mounted under the shelving, above the cultures. Although this has the serious disadvantage that the cultures on the shelves immediately above the lamps have a higher temperature than those underneath, it has the

116

advantage that it provides a more uniform irradiation of the cultures than in the second method of installation. In this method the lamps are mounted centrally between the shelves (See Fig. 13.1) giving less even light but presenting fewer problems with heating of the cultures. In both systems the problem remains that the culture vessels nearer the fluorescent tubes have a higher temperature than those further away. To ensure that the temperature variation between the shelves is as small as possible it is best to use open metal racks rather than boxed in structures. By increasing air movement within the chamber it is possible to partially prevent the increase in temperature within glass or plastic containers upon irradiation.

There are still three aspects of irradiation to be considered.

1. If the temperature in the growth chamber increases as a result of the use of fluorescent tubes, at a particular pre-set temperature the cooling system comes into operation. If for some reason the cooling system comes interrupted (e.g. leakage of the cooling system), then the temperature will rise until all the cultures are killed. To overcome this problem it is important that each growth chamber has a back-up safety system, so that the lighting cuts out when a particular critical temperature is reached.
2. When installing the fluorescent tubes, advice on the type of tube etc. should be obtained from the supplier, since this can have a considerable influence on the growth and morphogenesis of the plants. The fluorescent tubes need only be replaced when they are wornout.
3. We have established that the light output from fluorescent tubes declines from the moment they are switched on: if the amount of irradiation at the beginning of use is considered as 100%, then this falls to 93% within 8 days, to 86% after 4 months and to 70% after 12 months.

It is extremely difficult to indicate whether a particular culture should be grown in the light or the dark, at high or low temperature. It is usually best to choose those light and temperature conditions that are best for the growth and development of the experimental material in vivo. Sometimes when growing an embryo, a tissue, or an organ, etc. special conditions become necessary, first in the light and then in the dark or vice versa. For instance, a dark period followed by a light period is necessary for the formation of adventitious shoots from a young flower bud of *Freesia* (Pierik and Steegmans, 1975b). Sometimes light and sometimes darkness is needed for the germination of seeds, the seeds being known as light germinators or dark germinators respectively. The literature often gives opposing conclusions; one author reporting the need for light for organogenesis when working with plant X, and another author working with plant Y reporting the opposite.

13.2. Discussion of special physical factors

Little is known about the influence of physical factors because:

1. Experiments in which light and temperature are varied are relatively expensive. Many growth cabinets are needed with different temperatures, day-lengths, and light sources. It is possible to study the influence of irradiance using a simple experimental set-up as shown in Fig. 13.2. The irradiance can be reduced by covering the cultures with additional layers of cheese-cloth.
2. There are no experimental results showing the significance of feeding CO_2 and O_2 on in vitro culture.

Fig. 13.2. To study the effect of irradiance, the cultures can be placed on a step system built on a table in the culture room. The fluorescent tubes are mounted 150 cm above the table.

3. There is little available information concerning the microclimate in the test tube or flask. If the relative concentration of CO_2 in the tube is high, then there is little point in continuing with further addition of CO_2.

A short review of some of the known facts concerning the influence of physical factors on in vitro culture are given below.

Light. Light is a more complicated factor than was first thought; as well as day-length, irradiance and spectral composition (strongly dependant on the type of lamp) are of special importance. These three factors are considered later.

Little is known about the effect of day-length on in vitro cultures; the usual day-length chosen is 14–16 h, although continuous light is also used. In extremely special cases growth takes place in continuous darkness (dark requiring orchid seeds). If processes are being studied which are extremely sensitive to day-length (Fig. 13.3), such as dormancy and flowering, then care must be taken to choose from a particular range of day-lengths. In principle, the best day-length in vitro will be the same as used for cuttings or intact plants.

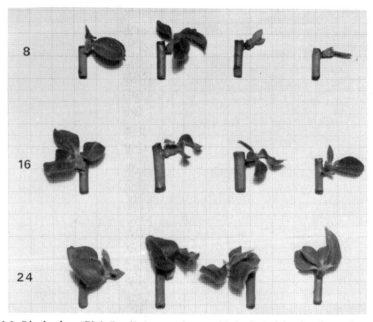

Fig. 13.3. Rhodendron 'Pink Pearl' stem explants, with buds (originating from plants which have been grown under long-day conditions), react favourably in vitro to day-lengthening from 8 to 16 or even to 24 h by outgrowth of the buds (Pierik, 1976). Photograph taken 6 weeks after isolation.

It is somewhat easier to give advice concerning the irradiance, since more is known concerning this factor. High irradiances, which prevail in the open field or in the phytotron (30–70 W m^{-2}), are almost without exception damaging for in vitro cultures. This can be explained by the high temperatures (which result from irradiation; the so-called 'greenhouse' effect), which are produced in test tubes and containers. Growth usually takes place at an irradiance of 8–15 W m^{-2}, or sometimes at very low irradiance (Fig. 13.4). A low irradiance is often chosen since it is assumed that photosynthesis in vitro is often limited by the low CO_2 concentration above the agar (Hanson and Edelman, 1972). In fact this supposition is not correct, since there is in fact a surplus rather than a shortage of CO_2 in the tubes and containers etc. If the tissues do not contain any chlorophyll then there is no point in irradiation (Bergmann, 1967). In the case of woody species there are indications that a higher irradiance than that advised (8–15 W m^{-2}) might be advantageous. Evers (1984) showed that with isolated shoot primordia of *Pseudotsuga menziesii* the irradiances which gave the same optimal effects on growth and development were 36.4 W m^{-2} when the isolation took place in January and 24.5 W m^{-2} in April. He also showed that too low an irradiance could not be compensated for by the addition of sugars. In contrast to these observations Pierik et al. (1986), showed that with *Syringa vulgaris*

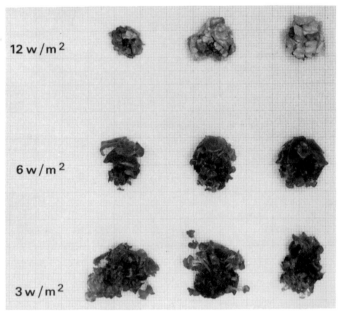

12 w/m^2

6 w/m^2

3 w/m^2

Fig. 13.4. Callus tissues of *Begonia venosa*, which regenerate adventitious shoots are very sensitive to irradiance. A low irradiance level (3 W m^{-2}) is optimal (Pierik and Tetteroo, 1986). Photograph taken 12 weeks after isolation.

120

the optimal growth and development takes place at a very low irradiance of 4–5 W m^{-2}. The supposition that plants in vitro become 'lazy' through an excessive sugar supply from the medium has certainly not yet been verified.

Fluorescent tubes (cool white type) are nearly always used for in vitro culture, although in a few cases it has been determined that better results are obtained under high pressure sodium light (Norton and Norton, 1986a). Dutch laboratories mainly use Philips fluorescent tubes, type 33 (38 W). The important influence that the choice of lamps can make has been reported by Schneider-Moldricks (1983). She studied the influence of 6 different types of fluorescent tubes on the adventitious shoot formation on *Kalanchoë* leaf explants. Best results were obtained with the types Interna 39 and Warm White de Lux 32, which are the lamps with a relatively high proportion of orange-red light. Lamps which gave out a relatively large proportion of UV or near UV light had an inhibitory effect on adventitious shoot formation.

Other work has also stressed the fact that the importance of the light source should not be under-estimated. Beauchesne et al. (1970), showed that with the addition of IAA red and green light were far more effective than blue light in stimulating the growth of tobacco tissues. It was later discovered that blue light breaks down IAA (although not NAA), resulting in the limitation of growth. These results do not hold true for poplar, willow, chrysanthemum and dahlia tissues. Red light (660 nm) also stimulates adventitious root formation in *Helianthus tuberosus* more than blue light (Letouzé and Beauchesne, 1969). Ward and Vance (1968) showed that growth of *Pelargonium* callus was better in white (polychromatic) and blue light in comparison to in green or red light or total darkness. Seibert et al. (1975) also reported that blue and violet light stimulated adventitious shoot formation in tobacco callus, whereas red light induced adventitious root formation. It can be generally concluded that white light usually inhibits adventitious root formation (Pierik and Steegmans, 1975a), and promotes adventitious shoot formation.

The following observation can be made about the different roles of red and infra-red light. In most cases it has been established that red light promotes adventitious shoot formation: *Pseudotsuga menziesii* (Kadkade and Jobsen, 1978), *Petunia hybrida* (Economou and Read, 1986), and *Brassica oleracea 'Botrytis'* (Bagga et al., 1986). Degani and Smith (1986), working with tobacco are the only people to report an inhibiting effect of red light on organogenesis. Red light has also been found to promote adventitious root formation in *Rhododendron* (Economou, 1986), and *Petunia hybrida* (Economou and Read, 1986). All these informations lead to the conclusion that the red/far-red system is important in the formation of adventitious shoots and roots. A detailed review of the role of light (wavelength, day-length, irradiance), can be found in the handbook by George and Sherrington (1984).

Temperature. The temperature is usually kept constant at 24–26 °C. Sometimes, depending on the experimental species a lower temperature (18 °C for bulbous species), or higher temperature (28–29 °C for tropical species; Fig. 13.5) is chosen. The optimal temperature for in vitro growth and development is generally 3–4 °C higher than in vivo. Since the temperature within the test tubes is 3–4 °C higher than the growth room, due to the warming effect of the irradiation, the growth room temperature can be maintained at the optimal for in vivo plant growth.

A lower temperature has to be chosen for some special processes, such as flower bud formation, breaking of dormancy, seed germination etc. Good in vivo growth and dormancy breaking are obtained with in virtro formed adventitious bulbs of gladioli, lily and *Nerine* species, when they are given 4–6 weeks at 5 °C. Sometimes very special conditions are needed for organ formation (Pierik, 1967): 5 °C for adventitous root formation of the biennial honesty, 27–29 °C for optimal adventitious root formation in *Gerbera* (Fig. 13.6), and 17–25 °C for adventitious bulb for-

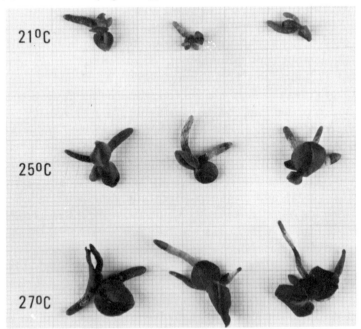

Fig. 13.5. The growth of *Phalaenopsis* hybr. seedlings in vitro is optimal at a temperature of 27 °C (Pierik and Hendriks, unpublished results). Photograph taken 14 weeks after isolation.

mation in hyacinth. Low or very low temperature can be used to stop the growth of in vitro cultures (freezing). More information on this can be found in Section 25.2.5.

Sometimes alternating temperature conditions may be needed; root formation in *Helianthus tuberosus* is promoted by a day temperature of 26 °C and a night temperature of 15 °C. De Capite (1955) came to the same conclusion with callus tissues of *Helianthus tuberosus,* carrot, and *Parthenocissus,* where the best growth was under a day temperature of 26 °C and a night temperature of 20 °C. When using an alternating temperature it must be borne in mind that a lowering of the temperature will result in an in-flow of air (air in the chamber will contract when cold), with a resulting increase in the number of infections, especially when using stoppered tubes or flasks.

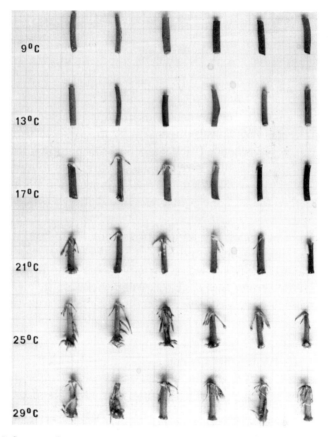

Fig. 13.6. Influence of temperature on regeneration of adventitious roots from apolarly placed petiole explants of *Gerbera jamesonii.* At low temperatures (9° and 13 °C) no root formation occurs, at higher temperatures (17°-29 °C) adventitious root formation is promoted (Pierik and Segers, 1973). Photograph taken 8 weeks after isolation.

123

Humidity. Little is known about the influence of the humidity of the growth chamber on in vitro growth and development. Since the humidity in the test tubes is relatively high, (as can be seen by the condensation inside the tubes), the humidity of the growth chamber probably only influences loss of water from the test tubes. However, a high humidity in the growth chamber results in a higher portion of infections. Information on the water loss due to evaporation, and the influence of the closure of the growth vessels on this is given in Chapter 7.

Availability of water. As was discussed in Chapter 6 on nutrient media, the availability of water (which can e.g. be controlled by the agar concentration) influences the chances of vitrification. Bouniols (1974) and Bouniols and Margara (1968) showed that with chicory explants the chances of flowering ware dependant on the availability of water: the flower bud differentiation is determined by the state of hydration of the chicory tissues. The production of flower buds is good on solid media and poor on liquid media; although it is possible in liquid media by lowering the humidity of the air.

Oxygen. Good aeration appears to be an important factor for the growth of cells, tissues, etc. This is well illustrated by the frequent use of shaking machines and orchid wheels, fitted with flasks containing liquid media. The oxygen supply in the test tube can be promoted as follows: only use metal çaps, not cotton wool plugs; use apolar inoculation (more O_2 available above rather than in the agar); use liquid media (better O_2 transport); inoculate on a paper bridge.

Organ formation, especially adventitious root formation is promoted by a better O_2 supply (Pierik and Steegmans, 1975a); it is a well established fact that especially cuttings of woody species find it extremely difficult to regenerate roots at their bases when these are in the agar. Root formation is much better in a liquid medium or in an well aerated substrate in vivo (better O_2 supply).

Carbon dioxyde. Although CO_2 can in principle be used as the carbon source for in vitro culture, in fact saccharose is a far better carbon source (Bergmann, 1967). It is certainly true that the addition of CO_2 in vitro serves very little purpose, since the CO_2 concentration in well sealed tubes or flasks is often already very high. More details on CO_2 feeding or enrichment can be found in the article by Chandler et al. (1972), and Gathercole et al. (1976). However, it must be taken into account that photosynthesis in vitro is often less than normal due to the low irradiance (Hanson and Edelman, 1972; Evers, 1984), in which case there is little point to CO_2 feeding.

Electric current. When a weak current, in the order of 1 µA is passed between the tissue and the culture medium, dramatic increases in growth of tobacco callus have been recorded (Rathore and Goldsworthy, 1985). This effect is dependant on the direction of the current; when the callus is made negative, the growth rate is increased by about 70%, whereas if the current is reversed there is only slight promotion.

14. The transfer from nutrient medium to soil

A plant which has originated in vitro, differs in many respects from one produced in vivo (de Fossard, 1977; Grout, 1975; Grout and Aston, 1977, 1978; Grout and Crisp, 1977; Debergh, 1982; Wardle et al., 1983; Sutter, 1985; Fabbri et al., 1986; Connor and Thomas, 1981; Brainerd and Fuchigami, 1985).

1. With plants grown in test tubes the cuticle (wax layer) is often poorly developed, because the relative humidity if often 90–100% in vitro. This results in extra water loss through cuticular evaporation, when the plant is transferred to soil, since the humidity of the air in vivo is much lower. Leaves of an in vitro plant, often thin, soft, and photosynthetically not very active, are not well adapted for the in vivo climate. Test tube plants have smaller and fewer palissade cells to use light effectively, and have larger mesophyll air space. Stomata do not operate properly in tissue culture plants; open stomata in tissue culture plants cause the most significant water stress during the first few hours of acclimatization. In tissue culture plants poor vascular connections between the shoots and roots may reduce water conduction. It must also be realized that the in vitro plant has been raised as a heterotroph, while it must be autotrophic in vivo: sugar must be replaced through photosynthesis.

It can be seen from the observations above that in vitro plants should be given time to get used to the in vivo climate and/or allowed to acclimatize (already in vitro), and become hardened off. Acclimatization can take place by allowing the in vitro plants to gradually get used to a lower relative humidity, which is the case in vivo. Development of a stomatal closure mechanism is a very important component of acclimatization. Wardle et al. (1983), have shown with *Brassica oleracea 'Botrytis'* that lowering of the relative humidity in vitro results in better wax formation on the cuticular layer, leading to less

127

cuticular evaporation. Acclimatization directly after the in vitro phase can be brought about by keeping the relative humidity high in vivo (Fig. 14.1), and maintaining a low irradiance and temperature. Another method of acclimatization is to leave the tube or flask open in a sterile environment for a few days to adjust to in vivo conditions. It is also possible to spray the plants with anti-transpirants to reduce evaporation in vivo (Sutter and Hutzell, 1985), although this often has an adverse effect.

2. Roots that have originated in vitro appear to be vulnerable and not to function properly in vivo (few or no root hairs); they quickly die off and must be replaced by newly formed subterranean roots. The development of root hairs in vitro can sometimes be promoted by allowing them to develop in a liquid medium. The poorly developed root system makes in vivo growth for such a plant very difficult, especially when there is high evaporation. It is vital that the in vitro plant looses as little water as possible in vivo (see the conditions discussed in point 1).

3. Plants which under normal growth conditions live symbiotically with fungi (mycorrhiza), or bacteria *(Leguminosae: Rhizobium),* lack these symbiotic organisms when they are transferred from test tubes to soil. There are progressively more reports that inoculation of in vitro grown plants with fungi or bacteria stimulates growth and develop-

Fig. 14.1. Young *Anthurium andreanum* plantlets, which after growth in a test tube have been transferred to sterile, sieved soil. To facilitate rooring in the soil the plants are covered, for a period, with beakers which helps to prevent evaporation (Pierik et al., 1975a).

ment. Dhawan et al. (1986) demonstrated that the addition of *Rhizobium* during hardening of *Leucaena* plants resulted in nodulation of 80% of the plants that survived transplantation to soil, with higher frequencies (90%) than the non-nodulated plants. Morandi et al. (1979) found that in vitro produced plants of *Rubus idaeus* grew much better when they were inoculated with a few *Glomus sp.* *mycorrhiza*. Strullu et al. (1984), reported that mycorrhiza play an important role in the in vitro propagation of birch plants: inoculation with the mycorrhiza *Paxillus involutus* resulted in a 75% promotion in growth as compared with non-inoculated plants. Recently Tremblay et al. (1986), discussed the use of actinorhizal and mycorrhizal symbiosis in the micro-propagation of alder plants.

When considering the problems associated with the non-functional root system it appears that special measures have to be taken into account, and a summary of these is given below:

— Allow at least root primordia to develop in vitro; these can then grow into proper subterranean roots in vivo.
— Where possible allow the conditions for development of roots in vitro (for instance in liquid media), which will then be functional in soil.
— Transfer the complete rooting phase to the soil or other substrate (Debergh, 1982). Dip the shoots in an auxin solution just before the rooting phase to encourage root formation. The work of Debergh (1982) has shown that these methods are not suitable for all species; in vitro shoots often remain very tender and soft.

Even if the different rules for transplanting from a test tube to soil or another substrate are followed the following should also be borne in mind:

1. To avoid infections by fungi and bacteria:
   — the agar (with sugar) should be well rinsed.
   — sterilized soil should be used, soil can be sterilized with steam or very efficiently by gamma radiation. However, in practice 'really' sterile soil is seldom used.
2. Immediately after transfer to soil all pathogens of the plants (slugs, insects, bacteria, fungi, etc.) should be eliminated, as the vitro plant is often weak.
3. Fungi such as *Fusarium* and *Pythium* can be treated with 0.15–0.25% Previcur-N (Schering), immediately after transfer from the test tube (Zimmer et al., 1981).

4. To avoid damaging the roots it is best to plant out in finely sieved soil.

5. To improve the establishment of in vitro plants in vivo, rooting should take place on a medium poor in salts (Murashige and Skoog, 1962, at half strength or Knop, 1884).

6. Sometimes it is necessary to give a cold treatment (4–8 weeks 5 °C) in vitro or immediately upon transfer in vivo, to break dormancy. The breaking of dormancy is often necessary with bulbs formed in vitro, and sometimes with shrubs and treees.

7. Plants which forms corms or bulbs should be transferred in this form from the test tube into the soil; then the chance of survival is better. Acclimatization and growth (shooting) of tuberous and bulbous species when transferred to soil, is better with larger corms and bulbs.

8. Acclimatization in vitro, especially by exposing the plants to reduced relative humidity increases the survival rate when the plants are transferred to soil (Ziv, 1986; Short and Roberts, 1986). If in vitro plants are transferred to a greenhouse, then both the relative humidity and irradiance should be gradually lowered (Ziv, 1986). Hardening-off procedures to increase the relative humidity in vivo were described by Griffis et al. (1983): humidity tents (with and without mist system), automatic mist systems (with or without plastic covering), plastic covers and fog systems.

9. Desjardins et al. (1986) demonstrated with strawberry plants that the acclimatization of in vitro produced plants in soil, is also promoted by a CO_2 enriched environment, and supplementary lighting, and this was verified by Lakso et al. (1986) using grapevines. Grout et al. (1986) tried to develop fully autotrophic plants in vitro with a positive carbon balance (by transferring defoliated plantlets to sugar free media); this technique was also applied to roses by Langford and Wainwright (1986).

A review of recent developments in plant tissue culture industry concerning transfer from the test tube to soil are given below:

1. Transfer of in vitro rooted plants to soil does not take place commercially as was given in Fig. 14.1, but in a large unit: squares of soil, covered in plastic which keeps the air humidity high (sometimes spray nozzles are used), and the temperature and irradiance low (shading) (Fig. 14.2).

2. Shoot cuttings, formed in vitro, of some species *(Gerbera* and *Rhododendron)* are transferred directly into an artificial substrate. Recently Twyford Laboratories have begun making a plug plate under

Fig. 14.2. When in vitro produced plants are transferred to soil, dehydration as a result of too much evaporation should be avoided. The humidity of the air can be kept high by covering the plants with plastic sheeting.

licence. This is a plastic plate with 400 holes which are filled with a rooting substrate; this swells after wetting and the shoots are held fast. Rooting in plug plates takes place in a special semi-sterile rooting room (climate chamber), in which the humidity of the air is kept high by spray nozzles. As soon as the root formation is sufficient and the shoots begin to grow, the cuttings are transferred on the plug plate to the greenhouse. Each individual cutting can be removed from the plug plate and grown on in soil.

Woody species are also now more frequently rooted in vivo rather than in vitro, since this saves labour. *Rhododendron, Kalmia, Amelanchier, Betula,* apple root stocks, *Vaccinium* and *Syringa vulgaris* (Zimmerman, 1985), are all woody species for which this method is used commercially.

3. The company Milcap France S.A. (Chemin de Montbault, Nuaillé 49340 Trementine, France), has produced a synthetic rooting substrate which has a polypropylene base (capable of being autoclaved, biologically stable, chemically inert, well aerated, with a capillary network). This substrate is available as flakes, which are suitable for in vitro rooting. For direct root formation in vivo there are also small blocks available in which the cuttings can be placed directly. This new

rooting substrate is highly suitable for rooting shoots in vitro (flakes), and in vivo (blocks).

4. McCowan (1986) has recently described the different possible methods of micro-propagation (a field developing very rapidly at the moment), with transplant plug systems. He also describes the necessary criteria for an ideal plus system.

5. In recent years a mist system has become more generally used for acclimatization.

On the basis of the above review of development, we can expect, in the very near future more new substrates to come on the market, which will enable the taking of shoot cuttings in vitro and in vivo to be far more efficient than was previously possible. The direct taking of cuttings in vivo is especially important. The price of in vitro plants will certainly fall (in vitro rooting is still relatively expensive), resulting in in vitro propagation becoming even more popular; specially equipped growth rooms will be needed for in vivo cuttings on artificial substrates.

If the in vitro produced plants need to be transported over long distances they have to be packed in damp cotton wool and kept in plastic bags. The temperature has to be well regulated during the transportation, especially when considering tropical species such as the oil palm which cannot tolerate a temperature of 15 °C, a minimum of being 21 °C being acceptable.

15. Aids to study

15.1. Literature study

Invaluable help for in vitro cultivation of higher plants, can be obtained from handbooks, congress and symposium abstracts and reports, and articles in journals. The most important handbooks and congress reports on the subject of in vitro culture are given in the literature list.

Since it is almost impossible to give all the names of the many journals which are relevant for plant tissue culturists, a list is given below of journals where the most important information on in vitro tissue culture can be found. This list is from a tissue culture bibliography (Pierik, 1979).

Acta Botanica Sinica
Acta Horticulturae
American Journal of Botany
American Orchid Society Bulletin
Annals of Botany
Canadian Journal of Botany
Comptes Rendus Academie Sciences Paris
Current Science
HortScience
In Vitro
Journal Japanese Society of Horticultural Science
Physiologia Plantarum
Phytomorphology
Plant Cell Tissue Organ Culture
Planta
Plant Physiology
Plant Science Letters
(Continued as Plant Science)

Scientia Sinica
Zeitschrift Pflanzenphysiologie
(Continued as Journal Plant Physiology)

Bibliographies are also important for studying the literature; there are only 3 found on the subject of in vitro culture of higher plants: Brown and Summer (1975), Pierik (1979), and Bhojwani et al. (1986).

Considering the enormous flood of publications in the form of articles, it is practically impossble, even in a limited area, to gather and collate all the incoming information. It is far easier to remain on top of the situation with the help of a computer. Given below is a simplified short description of such a computer literature search.

If a regular, (usually monthly) up-date of particular literature is needed, then contact should be made with a large library or documentation centre, which has experience with such matters. The information specialist can then begin making an interest profile. This is facilitated with the help of key words, which are weighted with regard to their importance. A list is given below of key words which are relevant for the in vitro culture of higher plants; they cover the information given in this book and are used by the documentation/information centre PUDOC in Wageningen for a computer literature search. Each key word has a value depending on it's importance; the higher the value the more important the word. A few key words have negative values and their values are substracted if a number of word values are added together. A * after the key word implies a word function after which other things are allowed. By the use of special measures it is possible to pick out, for instance, in vitro culture of algae, fungi, bacteria, yeasts, or that of animal or human tissues.
Relevant key words:

10 Axenic*	10 Callus*
10 Crown gall*	10 Explant*
10 Mericlone*	10 Plantlet*
10 Protocorm*	8 Agar tube*
8 Excis*	8 Inoculat*
8 Test tube*	8 Vitro*
7 Androgen*	7 Culture*
7 Suspen*	7 Protoplas*
6 Embryo*	6 Media
6 Medium	5 Muta*
5 Sprout tip	4 Adventit*
4 Morphogen*	4 Regenerat*
4 Segment*	3 Cell*
3 Isolat*	3 Tissue*

2 Anther*	2 Bud
2 Buds	2 Continu*
2 Cultivat*	2 Differenti*
2 Meristem*	2 Multiplicat*
2 Pollen	2 Propagat*
2 Root*	2 Shoot*
2 Stem*	2 Tip
2 Virus*	-3 Virus Inoculat*
-5 Cellulo*	-5 Horticult*
-5 Viticult*	

How can a computer decide whether a particular publication is relevant to the subscriber or not? The title of the publication together with the key words and a summary, if available, are considered, and the positive values of any keywords found added together; negative key words being subtracted. This final value of all positive and negative key words must be above a pre-determined value (decided upon by the library and the subscriber), before the title of the article and in some cases the abstract is printed, or recorded on tape or floppy disc. In the key word profile for in vitro culture of higher plants the activation value is 10. A mechanised library or documentation centre, with literature searching facilities obtains all the relevant lists of literature (usually from reference journals) which can be computer analyzed, for new literature for a particular client. Costs of a computer literature search are dependent on the number of databases that are consulted and the number of literature cards retrieved. The costs may be lowered if more people subscribe to the same search. A list of the most important databases for the in vitro culture of higher plants is given below:

1. AGRIS. AGRINDEX in printed form. This is a bibliography compiled by the Food and Agricultural Organization (FAO), for agricultural scientists.
2. CAB. This is produced by the Commonwealth Agricultural Bureaux International, in England. It incorporates about 30 reference journals on agricultural topics.
3. AGRICOLA. This is a Bibliography of Agriculture copiled by the National Agricultural Library in Beltsville, Madison, U.S.A.
4. BIOSIS. Known as Biological Abstracts. This is a very large collection covering all aspects of biology. It is produced by the Bio Sciences Information Service in Philadelphia, U.S.A.

It is obvious that there will be a certain amount of overlap between 1-3, and if information is needed concerning the agricultural aspects of the in vitro culture of higher plants, it will be necessary to consult all 3. It is not possible to obtain one output without duplications from 1–3, since they are rival concerns. The collection in 4 differs in that it is more biological and less agriculturally orientated.

It should be borne in mind that despite all care being taken about 40–50% of the literature retrieved by the computer search will be irrelevant, due to the need for a broad search in order not to miss any references.

People who are interested in such computer literature searches, and have no such facilities in their own countries can obtain free information from: PUDOC, P.O. Box 4, 6700 AA Wageningen, The Netherlands (Telex 45015 blhwg nl).

In 1982 Tateno et al. in Japan developed a new database for plant cell and tissue culture; this consisted of about 2,000 articles in 1982. In recent years more companies are advertizing computer literature searches.

15.2. Societies and associations

Tissue culture societies were already in existence in different countries before the formation of the International Association for Plant Tissue Culture (I.A.P.T.C.), and organized local meetings for researchers and people from industrial concerns interested in this subject. In the Netherlands in 1963 a tissue culture club was initiatied to bring together people interested in plant cell and tissue culture.

The I.A.P.T.C. has an important international function. When the first newsletter appeared in 1971, there were already 648 members, but by 1986 this number has increased to 2894 from 72 different countries. In each of the countries involved, a so-called national correspondent acts as go-between for the local members and the I.A.P.T.C., and is responsible for the distribution of a newletter 4 times per year. The governing body of the I.A.P.T.C., which edits the newsletter, chooses a country at the International Congress on Plant Cell and Tissue Culture to produce the newsletter for a 4 year period. At the 6th congress the governing of the I.A.P.T.C. was handed over to The Netherlands from the U.S.A. The Netherlands will co-ordinate the working of the I.A.P.T.C. for the period 1986–1990, and will also organize the 7th International Congress in 1990.

Most people working with tissue culture in the different countries, will be familiar with who is representing their country within the I.A.P.T.C., how they can become a member, and will already be receiving the regular newletters. If anyone does not have this information the address of the

International Division is: Dr. A.J. Kool, Secretary I.A.P.T.C., Plant Biotechnology Division, Zaadunie Research, P.O. Box 26, 1600 AA Enkhuizen, The Netherlands.

To remain update with all the developments in the field of in vitro culture of higher plants it is very important to read the regular newletters of the I.A.P.T.C. They contain review articles; congress, conference, symposium and seminar reports; short reports on meetings; research reports; addresses and information over courses; an up to date list of members and their addresses; recently published books and journals; newly published articles, etc.

15.3. Laboratory notebook, photographs and slides

It is advisable to always keep an up to date laboratory notebook (a ring file is quite suitable). In this book all the experimental details are noted: number of the experiment, date that media were made, details of the media, sterilization procedure, growth conditions, number of infections, observations and end results. In this way all the important details are kept safely in case they are needed later.

It is also a good idea to regularly take photographs and/or slides of the cultures, especially at the end of any observations: in this way an objective, visual result of the experiment is always available. The photographs may be used later in publications and the slides for lectures or seminars.

Photography of in vitro cultures is often more difficult than it at first appears. There are two main problems: reflection from the glass and sometimes the small size of the cultures. Reflection from the glass can be overcome by removing the sample from the test tube before taking the photograph or slide. However, if this is not possible or desirable, then the following measures are recommended:

1. To lay the tube or flask diagonally pointing backwards not vertically upright. Reflection can also be lessened or eliminated by altering the lighting.
2. The use of a polaroid filter.
3. Placing a so-called ringflash around the top of the tube or flask.
4. Placing a cold light source in the form of a ring (used with a stereomicroscope), around the top of the tube or flask.

By the use of measures 3 and 4 (Staritsky, personal communication), it is possible to photograph shadow free, since the light comes from all sides. When photographing very small objects a close-up lens should be used, or a camera can be mounted on the microscope. To get a good idea of the size of the object, it can be photographed together with a calibration bar. When the photograph is taken outside the tube, it is advised to place the object on graphpaper (available in different colours).

16. Embryo culture

16.1. Introduction

Embryo culture is the sterile isolation and growth of an immature or mature embryo in vitro, with the goal of obtaining a viable plant (Fig. 16.1). The information dealt with in this Chapter is taken in part from the following review articles: Tukey (1935), Randolph and Cox (1943), Kruyt (1951), Rijven (1952), Rappaport (1954), Zagaja (1962), Sanders and Ziebur (1963), Narayanaswamy and Norstog (1964), Maheshwari and Rangaswamy (1965), Dégivry (1966), Bajaj and Bopp (1971), Theiler (1971), Beasly et al. (1974), Norstog (1975), Jensen (1976), Raghavan (1966, 1976a, 1977, 1980, 1986), Monnier (1980), Johri et al. (1982), Hu and Wang (1986).

In 1904 Hännig was the first research worker who obtained viable plants from in vitro isolated embryos of the *Cruciferae*. In 1924 Dietrich grew embryos of different plant species and tried to establish whether embryos could still germinate without the completion of the dormancy period; he spoke of 'künstliche Frühgeburt' (artificial premature birth) when such embryos developed into complete plants after isolation. Laibach (1925, 1929) isolated embryos of *Linum* which aborted in vivo. Tukey (1933a, b) succeeded in getting normal plants from thousands of abortive embryos of early ripening cultivars of different stone-fruits. Van Overbeek et al. (1941) discovered that coconut milk stimulated the growth of *Datura* embryos; as a result of this finding research into embryo culture was greatly stimulated. Randolph (1945) isolated *Iris* embryos to shorten the breeding cycle. In 1945 Cox and his coworkers grew embryos of different cultivars of cabbage to speed up seed germination. From 1945 onwards embryo culture has been increasingly utilized, particularly by plant breeders in interspecific breeding programmes.

In principle there are two types of embryo culture:

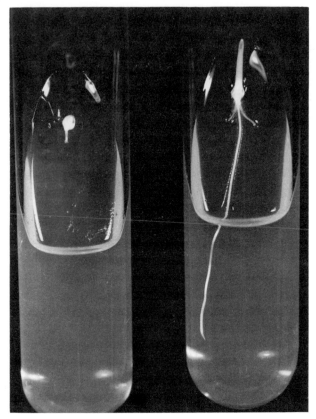

Fig. 16.1. Embryo culture of barley. Embryos were isolated on a solid medium slope. Left: 2 days after isolation, growth has just started (first root). Right: 6 days after isolation, a plantlet has developed.

1. Culture of immature embryos originating from unripe seeds. This type of embryo culture is mainly used to avoid embryo abortion (early death of the embryo), with the purpose to produce a viable plant. This type of culture is (extremely) difficult, not only due to the arduous dissection necessary, but also because a complex nutrient medium is needed. The chance of success of this type of culture depends strongly on the developmental stage of the embryo when it was isolated in vitro.

2. Culture of mature embryos derived from ripe seeds. This type of culture is relatively easy and is e.g. used to eliminate the (absolute) inhibition of seed germination. The use of a simple nutrient medium with agar, sugar and minerals for this type of culture is in most cases sufficient.

140

If the development of immature embryos in vitro and in vivo are compared starting from the globular stage then the embryos in vitro regularly have (Monnier, 1980):

1. A more bulky growth and are pear-shaped.
2. A retarded morphogenetic expression: a longer globular stage.
3. Initially only one cotyledon (in dicotyledonous plants) while two develop in vivo simultaneously.
4. The possibility of showing polycotyledonous development (more than two cotyledons), which seldom occurs in vivo.

In vitro isolated embryos usually exhibit precocious germination (Jensen, 1976), since there is a loss of inhibitors when the seed coat is removed; another reason might be that the (negative) osmotic potential in vivo has a much higher value than in vitro.

16.2. Techniques

With embryo culture there are normally no problems with disinfection. Single mature seeds are externally disinfected, then the embryo is removed after the seed coat is cut open. If the seeds are still unripe then the still closed fruit is disinfected and when this is opened the sterile ovules are available.

The dissection of the embryos produces more problems. It is possible to dissect large embryos without the use of a microscope, but this must be used, with a cold light source, for small embryos. Inoculation needles and empty holders with a piece of razor blade mounted at the end are used during the dissection procedures. Care should be taken when cutting the seed coat of a ripe seed as it is easy to damage the embryos. With seeds containing immature embryos it is usually easy to see where the small embryos are located and consequently there is less chance damaging them, although care should be taken not to damage the suspensor. A few examples of different isolation procedures are given below:

1. Isolation of cherry embryos is as follows (Theiler, 1971). First of all the stones are removed from the fruits; then the stones are thrown into a water bath to see which embryos are 'good' (those in the stones that sink in water). These are then disinfected. The sterile cherry stones are then cracked and the embryo can be seen. The embryos are then removed with the help of pointed forceps, and inoculated onto a solid medium.

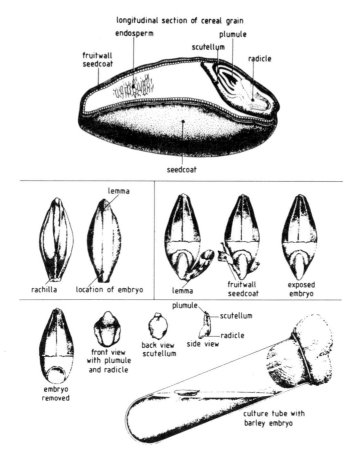

Fig. 16.2. Embryo culture of barley. For further explanation see text. Drawing from the Institute of Plant Breeding, Wageningen.

2. With barley the fruit is removed from the ear, disinfected, rinsed with sterile water and then laid in a sterile Petri dish with the rachilla underneath and the lemma above. After the removal of the lemma, fruit wall, and seed coat (Fig. 16.2), the embryo is free. It is then cut out with a knife and inoculated onto the medium (Lange, 1969). More details are given in Yeung et al. (1981).

3. With lily, 40–60 days after fertilization, the closed fruit is sterilized (96% alcohol and flamed), and then the seeds taken out and placed in sterile water to prevent dehydration (Smaal, 1980). The seed coat is removed by scraping with a knife and then the endosperm becomes visible. The embryo is isolated by lengthwise cuts, or it is pushed out of the endosperm through a hole made at one side. The embryo is inoculated as soon as it is free.

142

However, it cannot be said that the embryo culture is an easy process. In 1969 Lange isolated 1888 embryos of barley and only obtained 201 viable plants. The reasons for this low percentage viability are:

1. Loss due to infection (15%).
2. The embryos are too small when isolated (raising the sugar concentration promotes the growth of small embryos).
3. Interruption of the embryo development.
4. Damage (especially with hard seeds).
5. The artificial conditions (nutrient media).

Since embryos are sometimes difficult to dissect ovary and ovule culture may be used (Rédei and Rédei, 1955; Harberd, 1969; Braak, 1967, 1968). Braak proposed that ovary and ovule culture are suitable because:

1. The dissection of the embryos is often difficult (it is very easy to damage the embryo during isolation) and time consuming, requiring experienced personnel.
2. The medium needed for embryo culture is far more complicated than that needed for ovary and ovule culture.

Braak (1967) carried out successful ovule culture with tomato, *Rhododendron,* tulip, lettuce, lily, onion, *Hibiscus,* carnation and *Gerbera.* Bajaj and Bopp (1971) reported results with *Hevea, Papaver,* and *Nicotiana.* Further details on ovary, ovule, placenta attached ovule culture are reviewed by Collins et al. (1984).

Recently the more general term in ovulo embryo culture has been introduced in stead of ovule, ovary, and fertilized ovule culture. In ovulo culture is not likely to be useful where maternal tissue or the endosperm exerts an inhibitory action on the development of the embryo (Collins et al., 1984).

16.3. Factors affecting the success of embryo culture

The development of a viable plant from an embryo depends on many factors, which are briefly given below:

1. *Genotype.* In some plant species the embryos are easy to grow and some are difficult. There are even differences between cultivars of a given species.
2. *Developmental stage of the embryo at isolation.* Generally very small undifferentiated embryos are virtually impossible to grow in vitro

(Jensen, 1976; Monnier, 1980). The more developed the embryo in vivo the easier it is to culture it in vitro. Sometimes it is possible with the use of specialized techniques to culture very small embryos; a piece of endosperm from a mature seed (Ziebur and Brink, 1951; Williams, 1978) or a piece of hypocotyl tissue (Mott, 1981) is incubated in close contact with the embryo; the very young embryo can also be transplanted in the endosperm of a normal seed from the same plant species (Williams and De Lautour, 1980).

3. *Growth conditions of the mother plant.* Usually mother plants are grown in a greenhouse, although sometimes material is used, for example barley (Yeung et al., 1981), which has been grown in the field and then after cutting transferred to aerated nutrient solutions in a phytotron. Improving the growth of the mother plant under controlled conditions generally results in better endosperm development, and therefore better growth of the isolated embryos. The growth of the embryo is also promoted if the cotyledons are more developed (Kruyt, 1952; Kester, 1953), this also being dependent on the growth of the mother plant. In *Fraxinus* the seed dormancy is most probably located in the cotyledons; in this species the cotyledons should therefore be removed before embryo culture starts (Bulard and Monin, 1963).

Sometimes the plants (inflorescences) are first treated with gibberellins, before the embryos are isolated; this increases the size of the embryos and makes them easier to handle.

4. *Composition of the nutrient media.* Immature embryos call for a far more critical medium composition than required for mature embryos. However, both mature and immature embryos require macro- and micro-elements and sugar. Usually a solid medium with pH 5.0–6.0 is used. For embryo culture many different mineral media are utilized, which can be found in the literature (Pierik, 1979), Monnier (1976) developed a mineral medium for *Capsella bursa-pastoris,* which has also been succesfully used with other species. With peach it appears that the macro-salts of Knop (1884) stimulated the normal development of the embryo to rosette plants, while Murashige and Skoog medium (1962) resulted into abnormal plants (Toledo et al., 1980). The most important factors are ammonium and potassium ions (Jensen, 1976; Monnier, 1980).
Saccharose is generally used as sugar source, although glucose and fructose are sometimes suitable. Sugar is primarily important as an energy source, although it also has the role of lowering the (negative) osmotic potential of the nutrient media, especially with young embryos (Rijven, 1952). Mature embryos are usually grown on 2–3%

144

saccharose; while immature embryos thrive better on higher sugar concentrations (8–12%); the sugar demand decreases with the size of the embryo (Monnier, 1980). An agar concentration of 0.6–0.8% is usually chosen, higher concentrations resulting in the inhibition of growth (Stolz, 1971). Sometimes successful growth is achieved in a liquid medium (Hall, 1948; Rabechault et al., 1972).

Auxins and cytokinins are not generally used with embryo cultures, since they often induce callus formation (Monnier, 1976). Gibberellins sometimes have a promotive effect, especially when dormancy plays a role (Raghavan, 1980), and in some cases there is a vitamin requirement in embryo culture (Raghavan, 1980).

Although nowadays synthetic media are used, in older references the addition of such things as coconut milk, casein hydrolysate and malt extract are described. These complex mixtures are still suitable for the culture of immature embryos. It is assumed that especially a few amino acids in these mixtures are important as a nitrogen source (Raghavan, 1980); it appears that glutamine is especially necessary for the growth of immature embryos.

Since the nutrient requirements change during growth and development of the embryos, sub-culturing is sometimes necessary. Monnier (1976) described an ingenious system, in which 2 different media (one for mature and one for immature embryos) are present, next to each other, in the same Petri dish. The immature embryo is incubated onto the first medium, which gradually changes in composition due to diffusion of substances from the second medium; sub-culturing is no longer necessary using this system.

5. *Oxygen.* This is an important factor, and the oxygen requirement of embryo culture appears sometimes to be higher than the oxygen concentration normally present in air (Monnier, 1980).

6. *Light.* Sometimes isolated embryos need to be grown in darkness for 7–14 days, after which can be transferred to the light to allow chlorophyll formation (Smaal, 1980).

7. *Temperature.* The optimum temperature is dependent on the plant species used. Normally a relatively high temperature is used for growth (22–28 °C); although some species such as lily (Smaal, 1980) require a lower temperature (17 °C). A cold treatment (4 °C) might be necessary to break dormancy. With lily, it has been found that the plants produced have to stay for 4–6 weeks at a temperature of 5–9 °C to obtain good growth. Peach plants grow in the rosette form unless they have had a cold treatment (Toledo et al., 1980).

16.4. Practical applications

The most applications of embryo culture are given below:

1. *Elimination of the (absolute) inhibition of seed germination.* With a few species it is absolutely impossible to obtain germination in vivo, and in these cases embryo culture is essential: *Colocasia esculenta* (Raghavan, 1977), *Musa balbisiana* (Cox et al., 1960) and *Pinus armandii x P. koraiensis* (Bulard and Dégivry, 1965).
2. *Germination of seeds of obligatory parasites* without the host is impossible in vivo, but is achievable with embryo culture (Raghavan, 1977, 1980).
3. *Shortening of the breeding cycle* (Fig. 16.3). There are many species that exhibit seed dormancy which is often localized in the seed coat and/or in the endosperm. By removing these the seeds germinate immediately (Randolph and Cox, 1943; Rabechauld et al., 1969). Seeds sometimes take up water and O_2 very slowly or not at all through the seed coat, and so germinate very slowly if at all. In these cases, embryo culture offers a possibility to speed up germination. A few examples are: Brussels sprouts (Wilmar and Hellendoorn, 1968), rose (Lammerts, 1942, 1946; Flemion, 1948), apple (Nickell, 1951), oil palm (Rabechauld et al., 1968, 1969, 1972, 1973) and iris (Ran-

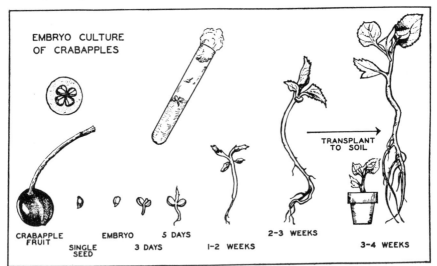

Fig. 16.3. Schematic illustration of the culture of an apple embryo. The fruit is sterilized, seeds are removed and the embryo is then isolated on a nutrient medium. After several weeks a viable seedling has developed, which can be transferred to soil. Since the seed coat impedes germination, this can be accelerated by embryo culture. Design after Nickell (1951).

146

dolph and Cox, 1943). It must be borne in mind that culture of immature embryos can result in an even greater shortening of the breeding cycle (e.g. in orchids, see Chapter 17).

4. *Production of haploids.* With the cross *Hordeum vulgare x H. bulbosum,* fertilization occurs but thereafter the chromosomes of *H. bulbosum* are eliminated. The result is that the haploid embryo of *H. vulgare* remains, which is only viable by embryo culture. After chromosome doubling a homozygote *H. vulgare* is finally produced (Kasha and Kao, 1970; Jensen, 1976). Following the *bulbosum* technique haploids were also obtained with *Agropyron tsukushiense* after crossing with *Hordeum vulgare* (Shigenobu and Sakamoto, 1981). Kott and Kasha (1985) have recently published a review of the uses of embryo culture and haploid plant production in cereals.

5. *Prevention of embryo abortion with early ripening stone-fruits.* In crosses of early ripening stone-fruits (peach, cherry, apricot, plum) the transport of water and nutrients to the yet immature embryo is sometimes cut off too soon, resulting in abortion of the embryo. In practice this means that no crossings with an early ripening mother are possible. The use of embryo culture to overcome abortion in vivo, is the only solution (Tukey, 1933a, b). It appears with peach that growing the embryos for as long as possible in vivo, is the best way to achieve success in vitro (Toledo et al., 1980).

6. *The prevention of embryo abortion as a result of incompatibility.* With interspecific crosses, intergeneric crosses and with crosses between e.g. diploids and tetraploids the endosperm often develops poorly or not at all. This results in embryo abortion in vivo, which can only be avoided by embryo culture. Embryo culture is frequently used with interspecific crosses: *Phaseolus* (Bajaj and Gosal, 1981), lily (Skirm, 1942; Smaal, 1980), flax, cotton, tomato, rice, and barley (Jensen, 1976). Well known examples of embryo cultures with intergeneric crosses are: *Hordeum x Secale,* and *Triticum x Secale* (Raghavan, 1977). Embryo culture is also used with crosses between diploids and tetraploids (resulting in triploids with very undeveloped endosperm): barley and rye.

7. *Vegetative propagation.* In, for example, *Gramineae* and *Coniferae* embryos are often used as a starting material for vegetative propagation. They appear to be very responsive because they are juvenile; with *Gramineae* organogenesis takes place relatively easily from juvenile callus tissue, while with conifers the cloning via immature calluses derived from young embryos and axillary shoot formation is also easy (Hu and Wang, 1986).

Extensive lists of the applications of embryo culture are given in the bibliographies by Pierik (1979), Narayanaswamy and Norstog (1964), Norstog (1975), Johri et al. (1982), Bhojwani et al. (1986) and in the review articles quoted at the beginning of Chapter 16.1

17. Germination of orchid seeds

17.1. Introduction

Bernard (1909) accidentally discovered that fungi are important for the germination of orchid seeds. It appears that orchids live in a symbiotic relationship with fungi, from the moment of germination. Symbiosis is the association of two organisms, to their mutual advantage. The symbiotic relationship between fungus and a root is known as a mycorrhiza; which literally means fungus-root. It was already established around 1900 that the hyphae (fungal threads) penetrate the protocorms and the roots of the plant and are 'digested' possibly making nutrients and other materials available to the orchids. It has been established that there are two different types of host cells in orchids; subepidermal host cells, in which the fungus can remain untouched and 'healthy', and more deeply lying host cells, in which the fungus is 'digested' into an amorphous mass (Hadley, 1975; Harley, 1969). The most important fungi that have symbiotic relationships with orchids are from the genus *Rhizoctonia*. Since in the early days (1900–1910), it was thought that orchids could only germinate in vitro in the presence of a fungus, only symbiotic germination or orchids was carried out. It was supposed that germinating and juvenile orchids were heterotrophic, and were reliant on the fungi for the supply of materials; the main reason for this assumption was the small amount of food reserve found in the orchid seeds.

Hadley (1982) concluded that it is often unclear what the orchid receives from the fungus. It is supposed to be carbohydrates, although this is far from certain in some cases. It has been established that the orchids sometimes receive amino acids (glutamine, glutamic acid, aspartic acid), and nicotinic acid from the fungus (Hyner and Arditti, 1973). In the case of tropical orchid germination it appears that the role of the fungus is not pronounced; whereas it is pronounced in terrestrial orchids.

Knudson (1922), showed that germination was possible on a simple medium containing minerals and sugars, in the absence of a fungus. The fungus is of little importance to the plant after the juvenile phase, since the plant is then a photosynthetic autotroph (Withner, 1974).

The research carried out by Knudson around 1922 was a shock to the orchid world, since he showed that seeds of *Cattleya, Laelia, Epidendrum,* and many other orchids were able to germinate asymbiotically in vitro. Despite Knudson's discovery it was soon realized that it was not often possible to make a medium on which a chosen orchid species could germinate and develop. Many nutrient media are described in the literature for many different genera and species (Arditti, 1967, 1982; Arditti and Ernst, 1982; Fast, 1980).

For the many outsiders who are not aware that most orchids are sown in vitro, the reasons for this are given below:

1. Orchid seeds are very small and contain very little or no food reserves. Their small size makes it very likely that they can be lost if sown in vivo, and the limited food reserves also make survival in vivo unlikely. Germination is much more successful in vitro.
2. As the germination and more importantly the further development of the seedling is dependent on a symbiotic relationship with a fungus in nature, then in principle, a fungus should also be present in vitro: this is referred to as a symbiotic germination. It is possible in vitro to be independent of the fungus by substituting it's action with a nutrient medium; this is known as asymbiotic germination.
3. If only a limited number of seeds are obtained as a result of a particular cross, then by choosing a suitable nutrient medium all of these can be germinated in vitro.
4. Sowing in vitro makes it possible to germinate immature orchid embryos; this leading to a shortening of the breeding cycle.
5. Germination and development take place much quicker in vitro, since there is a conditioned environment, and no competition with fungi or bacteria.

To have a better understanding of (the sowing of) orchid seeds articles by Arditti (1967), Arditti and Ernst (1982), and Harley (1969), should be consulted. Orchid seeds are very small; only 1.0–2.0 mm long and 0.5–1.0 mm wide. They are produced in large numbers: 1,300–4,000,000 per capsule. The seeds consist of a thickened *testa* (seed coat), enclosing an embryo of about 100 cells (Fig. 17.1). The seed coat has a characteristic net-like outer appearance which is species specific. The testa is dead tissue that is composed mainly of air (up to 96%): an orchid seed can then be seen to be like an air balloon. The embryo has a round or spherical

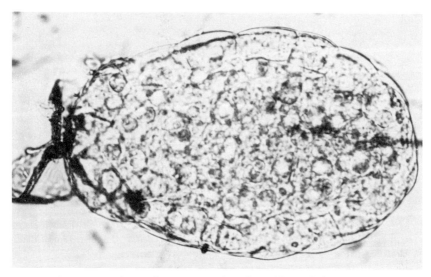

Fig. 17.1. The 'embryo' in a quiescent seed of *Paphiopedilum ciliolare*. The embryo is not differentiated and there are no cotyledons, roots or visible endosperm (Pierik et al., 1983).

form. Most orchid seeds are hardly differentiated at all: there are no cotyledons, no roots, and there is no endosperm. At the distal end of the embryo there is a potential growing point, which at this stage is not recognizable. The cells of the embryo have a simple structure, are isodiametric and poorly differentiated.

According to Arditti (1967), Harley (1969), and Pierik et al. (1982a, 1983), germination of orchid seeds is as follows. The embryo imbibes water via the *testa*, and becomes swollen. After cell division has begun

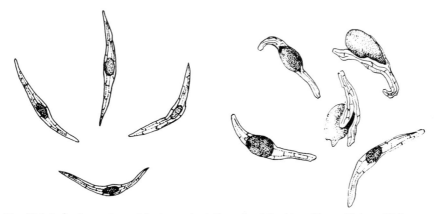

Fig. 17.2. Left: dry, quiescent (not germinated) seeds of *Paphiopedilum ciliolare*. Right: various stages of germination, 5–8 weeks after sowing in vitro (Pierik et al., 1983).

151

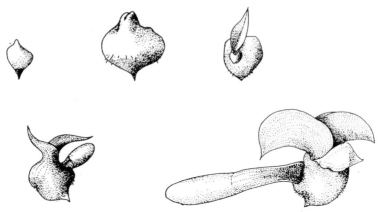

Fig. 17.3. The development of a protocorm of *Paphiopedilum ciliolare* into a plantlet. Above: the development of a shoot meristem into a shoot. Below: the formation of the first root. Drawings made 14–21 weeks after sowing in vitro (Pierik et al., 1983).

the embryo cracks out of the seed coat (Fig. 17.2). A protocorm-like structure is formed from the clump of cells, and on this a shoot meristem can be distinguished (Fig. 17.3). As soon as the organ differentiation begins (shoot meristem on one side and rhizoids on the opposite side), there is the period of profound growth. When in the light the protocorm becomes green and at the same time more leaves are produced. As a result of the chlorophyll formation the plant becomes an autotroph. Later the first real roots are formed endogenously; the protocorm and the rhizoids (root hairs) then loose their nutrient function and are lost (Fig. 17.3).

In relation to the germination behaviour, Hadley (1982) observed that epiphytic orchids generally germinated more quickly; the fungus growing on the outside. In contrast terrestrial orchids germinate slowly and the fungus penetrates into the cortex of the protocorm.

Until recently the percentage germination of orchid seeds could only be determined by sowing the seeds and counting those which germinated. However, Singh (1981), demonstrated that the tetrazolium test (viable embryos stain red, and non-viable ones remain white) can be used to determine the viability of seeds of tropical, epiphytic orchids. However, van Waes and Debergh (1986) have shown that Singh's method cannot be directly applied to West European orchids, the viability of which is best determined by a three step procedure: preparation of the seeds in a solution of 5% (w/v) $Ca(OCl)_2$ + 1% (v/v) Tween 80, followed by soaking in sterile water for 24 h, with a final application of the classical tetrazolium test.

Whether or not orchid seeds can be stored is difficult to answer; it is usually advised that ripened seeds should be stored at 5 °C (Fast, 1980).

Many research workers have reported that orchid seeds are already able to germinate before the capsule is ripe (has become brown), and has opened. This information is important for the amateur enthusiast as well as the professional worker (Valmayor and Sagawa, 1967), since:

1. It is far easier to sterilize a closed seed capsule, than individual seeds after they have come out.
2. Early propagation from the still green capsule is less exhausting to the plant.
3. There is less chance of embryo abortion with crosses between species and cultivars which are not closely related.
4. Sowing immature seeds shortens the breeding cycle; the seeds can sometimes be sown 2-3 months early without having to wait for the capsules to ripen and burst open naturally.
5. Sterilizing individual seeds is more likely to lead to damage.

Lucke (1971) claimed that, in principle, orchid seeds can be sown when the seed is half way between fertilization (this takes place very late with orchids), and ripening of the capsule. He advises that seeds should not be sown in vitro before the capsule is about two thirds ripe. In the table below (Lucke, 1971), are the normal ripening times of a capsule from the time of fertilization:

Calanthe	4	months	*Laelia*	9	months
Cattleya	11	months	*Miltonia*	9	months
Coelogyne	13	months	*Odontoglossum*	7	months
Cymbidium	10	months	*Paphiopedilum*	10	months
Cypripedium	3.5	months	*Phalaenopsis*	6	months
Dendrobium	12	months	*Stanhopea*	7	months
Epidendrum	3.5	months	*Vanda*	20	months

Sterilization can be carried out in two ways:

1. With the still closed capsule. In this case sterilization is basically simple: dipping in 96% alcohol and flaming. The seed capsule, which is usually internally sterile, is then opened with a sterile scalpel.
2. Opened capsules. Sterilizing individual seeds is a more complicated, and delicate procedure. The individual seeds are placed in a small flask (later closed with a rubber stopper), containing 2-5% (v/v)

bleach (10% NaClO), to which Tween 20 or 80 has been added. Sterilization takes 5–10 min. Viable seeds usually sink in bleach, and those that float generally contain no embryo. After sterilization the liquid is carefully poured or pipetted off, and the seeds are rinsed with sterilized water, etc. Alternatively the flask can be carefully turned upside down, with the stopper slightly loosened, the seeds mainly remaining on the sides of the flask.

Before the orchid seeds are sown it must be critically determined whether the seeds contain an embryo; this can be determined using a stereomicroscope under the correct lighting conditions. Innoculation is usually carried out on a sloping solid nutrient medium (which makes easier to spread the seeds). A spatula or inoculation needle is used to spread the seeds out evenly on the medium, or the seeds may be suspended in sterile water and inoculated in the form of a droplet (Arditti, 1982). The seeds are inoculated on and not in the agar where lack of O_2 can occur. The density at which the seeds are sown is dependent on the number of viable seeds (more dense sowing if viability is low), and also on the species of orchid under consideration. If *Phalaenopsis* is sown too thickly, the resulting protocorms and plantlets become inter-tangled and as a result pricking out becomes difficult.

It is debatable whether the seed coat (testa) contains inhibitors. Smith (1975) claimed that the seed coat has no inhibitory effect on germination. However, van der Kinderen (1982), proposed that the germination of certain orchids was influenced by the ABA-concentration of the seeds: high ABA-concentrations resulting in poor germination.

It should be appreciated that it may be necessary to prick out orchid seedlings in vitro 2–3 times, before they are big enough to be transplanted to soil. The medium into which they are pricked out is usually different to that on which they are sown. The length of the tube or flask phase is highly variable: 6 months to 2 years. Pricking out is necessary otherwise the flasks become over full and the growth stagnates. As is the case in vivo orchids in vitro are pricked out in groups; seedlings generally growing better in groups than singly, this being referred to in the literature as the 'community effect'. When the orchids are pricked out into soil they are usually transplanted into the so-called 'mother pot' (Fig. 17.4).

17.2. Factors affecting germination and growth

A large number of complex factors influence the germination and growth of orchids, these being highly dependant on the species. There follows a

Fig. 17.4. Paphiopedilum plants after their transfer to 'soil', growing together (community effect) in the so-called mother pot.

global appraisal of the different factors, although it is by no means complete. It has been compiled with the aid of the following articles: Arditti (1967, 1977, 1982), Arditti and Ernst (1982), Fast (1980), Hadley (1975, 1982), Harley (1969), Rao (1977), Thomale (1957), and Zimmer (1980).

When considering the list it should be borne in mind that the germination of immature seeds requires more regulated conditions than that of fully grown seeds; that terrestrial orchids are more difficult to germinate and grow than epiphytic (tropical) orchids. The specialized literature as found in e.g. the journals American Orchid Society Bulletin, die Orchidee, Orchideeën, etc. should be consulted when growing particular species.

1. Temperature. Seeds are usually germinated at 20–25 °C. Little is known about dormancy and possible cold requirement.
2. Light. A daylength of 12–16 h is usual (fluorescent tubes). A low irradiance of 2.5–10 W m^{-2} is advised. *Paphiopedilum ciliolare* seeds (Pierik et al., 1982a, 1983) only germinate when darkness is given for the first 3 months; this agrees the earlier research with *Paphiopedilum* carried out by Allenberg (1976), Smith (1975), and Stimart and

155

Ascher (1981). The related genus *Cypripedium* also requires darkness to germinate (Allenberg, 1976; Reyburn, 1978). After the seeds of *Paphiopedilum* and *Cyrpipedium* have germinated in darkness, a low light level is given (1 W m^{-2}), followed by an increase in irradiance (4 W m^{-2}).

3. Agar. Advised concentration of 0.6–0.8% (w/v Difco Bacto-agar).
4. Minerals. Although Knudson (1922) germinated orchid seeds on Knudson C medium (without micro-elements), it is advised that a medium is chosen with both macro- and micro-elements. Orchids often require iron to germinate (Fig. 17.5), and it is very probable that there is also a manganese requirement. The certainty of this

Fig. 17.5. Germination of seeds of *Paphiopedilum ciliolare* in vitro (Pierik et al., 1983). Left: germination on a medium without iron (poor germination, no further growth and development). Right: germination on a medium with 25 mg l^{-1} NaFeEDTA (good germination, growth and development). Photograph taken 28 weeks after sowing.

conclusion is questionable since agar is always contaminated with micro-elements (Fast, 1980).

Frequently used mineral media for the germination and growth of orchids are: Knudson C (1946), Vacin and Went (1949), Thomale GD1 (1957), Burgeff-3f (Thomale, 1957). Despite Thomale GD1 containing no calcium, this medium is extremely suitable for *Paphiopedilum* (Flamee, 1978; Stimart and Ascher, 1981; Pierik et al., 1982a, 1983). The mineral requirement of orchids is generally not high, and a salt poor medium is usually recommended. Arditti (1982) advised sowing all tropical orchids on a modified Knudson C medium. Terrestrial orchids require a far more complex salt poor medium, while epiphytic orchids need one that is richer in salts.

5. Sugar. This is extremely important as an energy source, especially for those that germinate in the darkness (these are completely heterotrophic as far as sugar is concerned). A sugar concentration of 1-3% (w/v) is used, generally saccharose although simetimes a mixture if glucose and fructose is used (both at concentrations of 0.5-1.5% w/v).

6. The pH must be chosen in the range 4.8-5.8, a pH lower than 4 or higher than 7 being extremely unfavourable.

7. Vitamins. The vitamin requirement is often disputed (Arditti, 1977a; Arditti and Harrison, 1977). The following vitamins are sometimes added: biotin, nicotinic acid (niacin), vitamin C, vitamin B_1, pyridoxin, panthothenic acid and meso-inositol. Nicotinic acid is most frequently used, although biotin is important for chlorophyll formation in *Paphiopedilum* (Lucke, 1969).

8. Regulators are usually not necessary for seed germination, and their addition often leads to unwanted effects (callus formation, adventitious shoot formation etc.). Low concentrations of IBA and NAA sometimes promote the growth of *Dendrobium, Cymbidium, Cattleya* and *Galeola* seedlings (Cheng-Haw et al., 1976–1978; Strauss and Reisinger, 1976; Nakamura, 1982). Cheng Haw et al. (1976–1978), also reported the promotive effect of kinetin with *Aranda* and of GA_3 with *Vanda*.

9. Complex mixtures. The following mixtures are used: banana homogenate, coconut milk, peptone, tryptone, brewers yeast, casein hydrolysate, salep, pineapple juice, tomato juice, potato extract (Arditti, 1982). The promotive effects of these mixtures is explained by their vitamin, amino acid, and sometimes regulator content. Peptone and tryptone (Fig. 17.6) have a promotive effect on the germination and growth of *Paphiopedilum* (Fast, 1971; Pierik et al., 1983). Ernst (1975) proposed that the addition of banana is especially promotive

Fig. 17.6. Effect of tryptone on germination and seedling growth of *Paphiopedilum ciliolare* in vitro. Left: without tryptone; right: with $2\,g\,l^{-1}$ tryptone (Pierik et al., 1983). Photograph taken 28 weeks after sowing.

for the growth of *Paphiopedilum* and *Phalaenopsis* seedlings; no marked differences were found between green and ripe bananas. The use of banana when sowing *Paphiopedilum* should be avoided, since this causes the growth after germination to stop completely (Pierik et al., unpublished results).

10. Active charcoal. Ernst (1975) and Fast (1971) found that the growth of *Paphiopedilum* and *Phalaenopsis* seedlings was promoted by 0.2% ($2\,g\,l^{-1}$) active charcoal; this in combination with banana is even more effective. On the contrary to what Ernst (1975) advised, the use of active charcoal when sowing *Paphiopedilum* should be avoided (Pierik et al., unpublished data).

18. Vegetative propagation of orchids

18.1. Introduction

Orchids are propagated vegetatively as well as generatively. With vegetative propagation, the progeny is identical to the parent plants. However, with generative propagation (by seed), identical progency are rarely obtained, and then only when it concerns the wild species. Therefore, if seeds from a cultivated orchid are used (mainly obtained from a cross and strongly heterozygous), the progeny will be extremely heterogeneous, seldom identical to the starting material. In principle, cultivated orchids can only be propagated vegetatively. Orchid cloning in vivo is a very slow process, requiring sometimes 10 years before a clone of some size is obtained.

After 1960 a revolution took place in the vegetative propagation of orchids. Morel (1960) attempted to obtain virus-free *Cymbidiums* by meristem culture; and for this isolated shoot tips in vitro, as shown in Fig. 18.1. 'Protocorm-like bodies' were obtained from these shoot tips, which were extremely similar to those already well known from seed germination. Sometimes such a protocorm divides spontaneously, but usually this is brought about by cutting the protocorm into pieces (Morel, 1965). It is possible with *Cymbidium* to produce 6–8 new protocorms from a piece of protocorm, in about 6 weeks; in principle, this cutting up of the protocorms can be repeated endlessly. When cutting is stopped, the shoot tip in each protocorm can, in principle, develop a shoot with leaves and roots (Fig. 18.2). In this way Morel produced thousands of plants in one year, beginning with only one shoot tip (a meristem with a few leaf primordia). Cloning of orchids by meristem culture, became the first commercial application of vegetative propagation in vitro. This method, originally developed for *Cymbidium* was later modified and used for *Cattleya*, and many other orchid genera. 'Meri-cloning' of orchids (vegetative propagation by meristem culture), is now carried out on a large

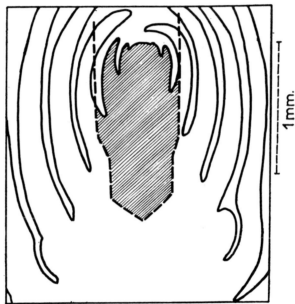

Fig. 18.1. Diagram of a longitudinal section through a *Cymbidium* bud. The meristem with 2 leaf primordia (shaded area) removed from the outer leaves, by cutting along the dotted lines, for isolation in vitro. Adapted from Champagnat et al. (1966).

scale world wide. As a result of this orchids are no longer luxury plants, indeed for some species such as *Cymbidium* too many plants are now produced.

A protocorm that has developed after germination from an orchid seed, has a morphological state that lies between an undifferentiated embryo and a shoot. With vegetative propagation by meristem culture (Fig. 18.2), protocorm-like bodies are also formed, which means in fact that a shoot tip of a adult plant rejuvenates (returns from the adult into the juvenile phase). Since protocorms obtained by seed germination have many close similarities (rhizoid formation, spontaneous or induced protocorm formation from pieces of protocorm, weak polarity), with those produced from isolated shoot tips, the term protocorm-like bodies has been introduced when cloning orchids by meristem culture (Champagnat, 1977).

The vegetative propagation of orchids by meristem culture can be divided into three phases: transformation of the meristem into a protocorm(-like body), the propagation of protocorms by cutting them into pieces, and the development of these protocorms to rooted shoots (Teo, 1976-1978). There is more detail on these different phases in Section 18.2 (Withner, 1974; Champagnat, 1977; Arditti, 1977; Arditti et al., 1982; Rao, 1977; Fast, 1980).

160

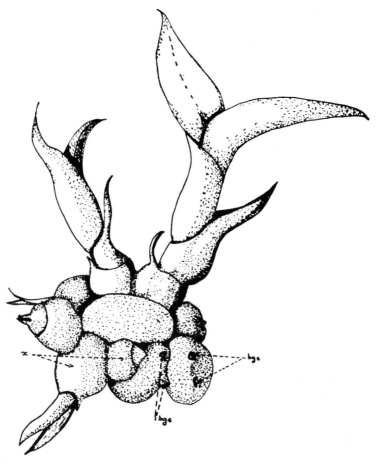

Fig. 18.2. Cymbidium meristem with several adventitious protocorm-like bodies and shoots. Adapted from Champagnat et al. (1966).

18.2. Meristem culture

18.2.1. *Production of virus-free plants*

One of the conditions necessary for propagation, is that the plantlets so produced should be virus-free. In Chapter 19 it is suggested that virus-free plants are automatically obtained with meristem culture, but in fact this is not so. The research by Morel (1960) with *Cymbidium* showed that virus-free plants were only obtained if the starting material was a meristem (of approximately 1 mm) with two leaf primordia. The cloning nowadays of far larger shoot tops, than used by Morel has the result that there

161

is little chance of obtaining virus-free plants. For propagation of orchids therefore virus-free plant material should first be obtained.

The most common virus found in orchids is a specialized strain of the tobacco mosaic virus (TMV-O), previously known as *Odontoglossum* ring spot virus. This virus is carried in the sap and contamination can take place from hands or a knife. It is the cause of ring-shaped spots on the leaves of *Cattleya* and also effects the form and colour of the flowers of many other orchids (Hakkaart and van Balen, 1980). A less common virus is the *Cymbidium* mosaic virus, also transferred in the sap, which is responsible for chlorotic patches or stripes on the leaves of *Cymbidium*, *Phalaenopsis* and *Vanda*, and necrosis of the flowers of *Cattleya*. Hakkaart en van Balen (1980), also propose the existence of the so-called 'Rhabdovirus' which attacks many orchids. In contrast to the first two viruses the Rhabdovirus is much more difficult to identify: an electron microscope is necessary, and a decision can only be made from the symptoms on a piece of an older leaf, which are not yet present on the in vitro propagated material.

18.2.2. *Propagation through meristem culture*

According to Champagnat (1977), Rao (1977) and Fast (1980) young growing shoots of 10–15 cm length which just have developed their leaves, should be used as the starting material for meristem culture. Any soil or debris is washed off with running water, and the leaves are removed until the axillary buds are visible. The leafless shoot is dipped in 70% (v/v) alcohol, and then sterilized for 10 min in 10% (v/v) bleach. The axillary buds are then prepared after they have been rinsed in sterile water. They are then sterilized again [10 min 3–10% (v/v) bleach, containing Tween 20 or 80], and re-rinsed with sterile water. The bases of the smallest buds are cut off in the laminar air-flow cabinet, and inoculation carried out. A few leaves are first removed from the larger buds the same as the bases which were killed by the bleach treatment, and then inoculation takes place. It takes a relatively long time after isolation (6–10 weeks), before the first protocorm-like bodies are formed. After this, the protocorms are cut up and sub-cultured etc. Nowadays, far larger buds (with at least 3–4 leaf primordia) are inoculated, this resulting in a more successful level of striking. If the protocorms are not cut, then the shoot tip on the protocorm grows out to form roots and shoots; if they are cut then the adventitious protocorm formation continues.

Champagnat (1977) considered that meristem propagation of orchids was not an accurate description of the process, since in fact the apical meristem plays no role and in fact is lost during culture. With meristem culture of *Cymbidium* only the zones around the leaf primordia show swellings, which ultimately are transformed into protocorms. Protocorm formation appears also to be possible without the apical meristem.

With meristem culture of *Cattleya*, browning quickly takes place on the wound tissues; for this reason, the advise is often given to cut the meristem in liquid, and to grow it in a liquid medium, in which the brown colouration more readily diffuses (Fast, 1980). A whole bud (3–5 mm) with many leaf primordia is usually isolated with *Cattleya*. The medium used for *Cattleya* is more complicated than that for *Cymbidium* (Fast, 1980; Champagnat, 1977); and sometimes contains auxin, cytokinin, coconut milk, and peptone. Protocorm formation with *Cattleya* takes a long time and always begins at the oldest leaf bases (Fig. 18.3); the apical

Fig. 18.3. Isolated meristem with leaf primordia of *Cattleya aurantiaca* 2 months after culture in a liquid medium. At the bases of the outside scale-like leaf primordia protocorms are formed on which shoot tips are already recognizable (Pierik and Steegmans, 1972).

meristem plays no role and is in fact lost. Isolated leaf primordia of *Cattleya* are also responsive and can form protocorms.

Miltonia, Odontonia, and *Odontoglossum* also appear to have the same response patterns as *Cattleya* (regeneration at the leaf bases), and are also grown on a relatively rich medium. The responses of these 3 genera are much more rapid than those of *Cattleya* (Champagnat, 1977). Up until now the meristem culture of *Paphiopedilum* has been unsuccessful, due to the rapid browning when the meristem (shoot tip) is isolated.

The medium used for meristem culture is very similar to the relatively simple sowing medium for the different genera. *Cymbidium* and *Miltonia* (relatively easy to propagate) are often grown on Knudson C (1946) or modified Vacin and Went (1949) media. *Cattleyas* and a few other genera need a special medium (Morel, 1974). The standard media for the different genera are given in Bergman (1972), Withner (1974), Arditti (1977), Fast (1980) and Zimmer (1980). The Murashige and Skoog (MS) medium (1962), and the Wimber medium (1963) were used later (de Bruyne and Deberg, 1974).

Isolation of meristems usually takes place in solid media, with the exception of *Cattleya*; protocorm propagation usually takes place in liquid media and the growth of the protocorms to plantlets again on solid media (Leffring, 1968). The composition of the media in the 3 different phases of orchid propagation is often different. It is striking that protocorms do not usually grow into plantlets on a liquid medium (Homes et al., 1972).

The pH of the media is chosen between 4.8 and 5.8. The sugar source is 1–3% (w/v) saccharose, or sometimes 1.5% glucose + 1.5% fructose. Monopodial orchids (Sagawa, personal communication) need a sugar-free medium for meristem isolation; in the presence of sugar yellowing takes place and the meristem dies. In a later phase (protocorm formation) sugar and even coconut milk may be added to the medium. In analogy to seedling culture coconut milk is often used (10–15%) to stimulate protocorm formation in the epidermal cells (Morel, 1974).

The addition of regulators is not generally necessary (Homes et al., 1972); the chances of mutation being greater in their presence. Positive effects of auxins were found by Fonnesbech (1972) with *Cymbidium,* and by Cheng-Haw et al. (1976–1978), with *Cattleya* and *Aranda*. A promotive effect of cytokinin was proposed by Reinert and Mohr (1967) and by Pierik and Steegmans (1972) with *Cattleya* and by Cheng-Haw et al. (1976–1978) with *Cattleya* and *Epidendrum*. Finally Fonnesbech (1972) proposed that GA_3 induced the elongation growth of *Cymbidium*.

The optimum temperature for orchid propagation lies between 22° and 28 °C. Fluorescent light is generally used with a daylength of 12–16 h, although sometimes continuous light may be used. A low irradiance can be used for the isolation, but this must be increased when plantlets are forming from the protocorms.

In principle the propagation of orchids by meristem culture can take place in two ways: 1. On solid medium; 2. In liquid medium which is kept in motion by a rotating wheel (Wimber, 1963). The speed is extremely variable, but most researchers prefer a slow rotation rate (2–5 rev. per min). Propagation and growth is usually better on a moving rather than a solid medium. There is a better O_2 supply in a liquid medium and the transport of nutrients is more effective.

When the plantlets, with about 3–4 leaves are big enough (5–7 cm high) after repeated pricking out, they can be transferred to boxes. It is advisable to sterilize the soil well (by steaming). The compost used is dependant on the orchid species being propagated. Soft peat is often used, the pH of which must be increased and nutrients added.

18.3. Other methods of propagation

Over the years a few other methods apart from meristem culture have been developed to clone orchids on a small scale:

1. Using young leaves. Young leaves and leaf tips of e.g. *Cattleya* form calluses at their bases from which protocorms can develop which are than available for further propagation (Champagnat, 1977; Champagnat and Morel, 1969). Churchill et al. (1971a, 1971b, 1973), also described the regeneration of plantlets, after callus and protocorm formation, from young leaves and leaf tips of *Laeliocattleya* and *Epidendrum*. Vij et al. (1984) propagated *Rhynchostylis retusa* by direct organogenesis from leaf segment cultures.
   Tanaka and Sakanishi (1977, 1985), proposed that protocorm-like bodies could develop on pieces of leaves from young shoots of *Phalaenopsis* under the influence of BA, NAA, and adenine, from which plantlets developed. The young shoots were obtained from isolated buds of an inflorescence (Fig. 18.4).
2. Dormant buds. Shoots can be formed from young primordial flower buds or dormant basal vegetative buds from the inflorescences of *Dendrobium, Vanda, Phalaenopsis, Oncidium, Phragmipedium, Epidendrum*, and *Thunia* (Rao, 1977; Fast, 1980; Singh and Prakash, 1984). According to Teo (1976–1978), Rotor, in 1949, was the first researcher to clone *Phalaenopsis* in this way.
3. Young inflorescences. Intuwong and Sagawa (1973), developed a method to propagate *Vascostylis, Neostylis*, and *Ascofinetia* in vitro

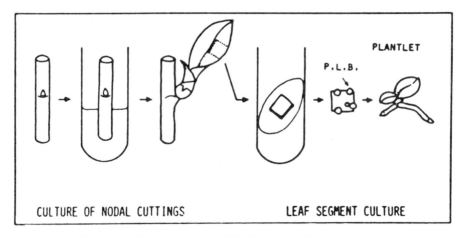

CULTURE OF NODAL CUTTINGS   LEAF SEGMENT CULTURE

Fig. 18.4. In vitro vegetative propagation of *Phalaenopsis* hybr. Leaf explants are isolated from shoots which have arisen from resting buds. These explants then form protocorm-like bodies (P.L.B.) which eventually produce plantlets (Tanaka and Sakanishi, 1977).

from explants of young inflorescences. These explants developed tissues which transformed into protocorms, when grown on a special medium.

4. Callus. Champagnat and Morel (1972), regenerated callus tissues from rhizome tissues of *Ophrys*, from which protocorms could be formed. The same was found to be true with *Aranda* (Chiang-Shiang et al., 1976–1978).
5. Young seedlings. These may be induced to regenerate as was shown by Pierik and Steegmans (1972) with *Cattleya aurantiaca*, where cytokinin was the inducing factor.
6. Etiolation. If in vitro grown seedlings of *Paphiopedilum* are etiolated by growth in darkness or at low irradiance, they can then be used for the single-node method (Pierik and Sprenkels, unpublished), which is described in Section 20.2, for plants other than orchids.

18.4. Variation arising during culture

Vajrabhaya (1977), ascribed the deviations which are encountered during propagation of orchids, to 3 causes:

1. Chimeras. Plants which are naturally chimeras, often have heterogeneous progeny when propagated in vitro; protocorm regeneration, however, is from one cell layer, which results in the loss of the chimeric characteristics. This is seen with *Cymbidium* and *Dendrobium*.

2. Viruses. As was described in Section 18.2, viruses can cause deviations, which resemble mutations.
3. Mutations. The most important cause of deviations is mutations, which are especially prevalent when regulators are used and/or when repeated propagation takes place. Vajrabhaya (1977), observed that with 205 plants from a clone of *Dendrobium*, 5 tetraploids (large flowers) were obtained. By in vitro propagation it appears that triploid *Dendrobiums* can become hexaploids; diploid *Dendrobiums*, after repeated propagation become octaploids. Rutkowsky (1983) showed that BA concentrations lower than $0.1 \, \text{mg} \, \text{l}^{-1}$ induced mutations in *Phalaenopsis*.

Mutations can also be induced artificially with colchicine or by ionizing radiation. Fast (1980) showed that the use of colchicine with *Dendrobium* (5 days, 0.1% w/v), and *Cymbidium* (10–21 days in liquid medium containing 2.5% colchicine) produced mutations.

18.5. Practical applications

There are many publications concerning the in vitro cloning of orchids, the most important of which can be found in a bibliography by Pierik (1979), and in the handbook by Arditti (1977).

19. Production of disease-free plants

19.1. Introduction

Internal infections caused by viruses, mycoplasmas, bacteria and fungi, can be very difficult to combat. It is virtually impossible to eliminate the plant of these internal pathogens by the use of chemicals. It is sometimes possible to suppress the virus multiplication by the use of the relatively expensive compounds Virazole (ribavirine) and vidarabine (antimetabolites), in the nutrient medium (Walkey, 1980; Kartha, 1986; Long and Cassells, 1986). In a few cases, such as with lily (Cohen, 1986) and apple (Hansen and Lane, 1985), the use of Virazole has resulted in the production of virus-free plants. The addition of antibiotics to counter internal bacterial infections is equally ineffective. In a review article Bastiaens et al. (1983) state that bactericides are of no use for eliminating bacteria in in vitro cultures, since the necessary concentrations are phytotoxic; sometimes the micro-organisms do not react to the added antibiotics at all. The use of antibiotics can also result unwittingly in the selection of resistant strains.

In contrast to what was thought previously, viruses can also be transferred during generative propagation; about 600 plant viruses are known of which at least 80 can be transferred in the seed (Phatak, 1974). Viruses, bacteria, and fungi are nearly always transferred by vegetative propagation. Of course an attempt can be made to sanitate a clone with the help of selection (by presence of symptoms, or by assay); however, if it appears that the clone concerned contains only infected plants, then an attempt must be made to rid them of the infection. Viruses especially, can result in the loss of plant production, and also be the cause of a poor quality product; so it is extremely important to use virus-free starting material when propagating vegetatively. Virus-free is used to mean free from those viruses which have been identified, and whose presence is discernable with e.g. a virus assay (Quak, 1966).

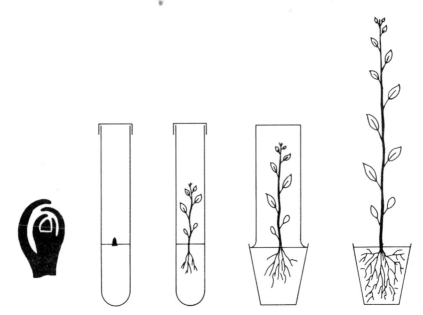

Fig. 19.1. Schematic survey of meristem culture. Left: a shoot tip from which a meristem is isolated in vitro. A plantlet develops from this meristem, which can be transplanted into soil. Drawing from the Institute of Phytopathological Research (IPO), Wageningen.

There are 5 available methods of producing virus-free plants: 1. heat treatment; 2. meristem culture (Fig. 19.1); 3. heat treatment followed by meristem culture; 4. adventitious shoot formation followed by meristem culture; 5. grafting of meristems on virus-free (seedling) root stocks; also known as micro-grafting.

To make large scale virus-free meristems available, in vitro adventitious shoot formation on explants may be induced, after which meristem culture is carried out. If isolated meristems do not take in vitro then these can be temporarily grafted on in vitro grown virus-free (seedling) rootstocks. Virus-free plants may also sometimes be obtained by callus or protoplast culture. Meristem culture is also often used to obtain plants free from bacteria or fungi. More about these methods of obtaining plants free from disease can be found in the following Section.

If it is suspected that a plant may be virus infected then the following procedures must be carried out (Quak, 1977):

1. Firstly the virus or viruses must be identified.
2. Secondly an attempt should be made to eliminate (all) virus(es).
3. The plants obtained should be assayed to see if they are virus-free.
4. Lastly any further re-infection should be prevented.

It must be borne in mind that virus-free material can be re-infected; the following procedures should prevent re-infection (Quak, 1966; Theiler, 1981):

1. The plants should be grown in greenhouses that are free of infections and vectors (apid free), or in a place where there are naturally no carriers of viruses.
2. Carriers of diseases (mainly insects and nematodes), should be controlled; sources of infection continuously removed.
3. Strict hygiene is necessary: disinfecting hands, clothes, shoes, as well as grafting and inoculating knives; prevention of mechanical transfer during the care of the crop.
4. The pots and substrates should be germ free.
5. For complete certainty (in case by mistake 1–4 are not carried out properly), it is recommended that disease-free plant material is maintained in vitro.
6. To carry out continuous selection, visually and by assays.

When obtaining virus-free plants it should be borne in mind that the infected plants may have a different appearance than those which are virus-free. Although this different external appearance is usually the result of virus elimination, micro-mutations cannot be ruled out as a possible cause. Walkey (1980), reported that micro-mutations occurred in rhubarb and apple after meristem culture. More about this phenomenon can be found in Chapter 21 on somaclonal variation.

The following articles have been consulted when writing this chapter: Morel (1964), Quak (1966, 1972, 1977), Styer and Chin (1983), Walkey (1980), and Hildebrandt (1977). More specialized information about particular genera and species can be found in the bibliographies by Pierik (1979), and Bhohwani et al. (1986).

19.2. Production of virus-free plants

19.2.1. *Heat treatment*

Heat treatment is an effective way of inactivating some viruses (Quak, 1972; ten Houten et al., 1968). It appears that heat treatment is only effective against isometric viruses and against diseases resulting from mycoplasmas (Quak, 1977). The fact that heat treatment is sometimes ineffective can be due to the plant itself being too sensitive to heat, or to the viruses or sometimes mycoplasmas being unaffected by heat for unknown reasons. A temperature and treatment time should be chosen which allow the plant (shoot, branch) to just survive, while the virus is inactivated.

Heat treatment is especially effective against viruses and mycoplasmas which are found in fruit trees, sugar cane and cassava (Morel, 1964; Quak, 1966, 1977; Kartha and Gamborg, 1975). This is extremely convenient in the case of fruit trees, where meristem culture is difficult.

In the case of woody species only the axillary buds rather than the whole plant are made virus-free with heat treatment. With these plants a branch which has been grafted onto a seedling is usually taken after it has been grown for 20–40 days (preferably a few months), at a constant or alternating temperature of 37–38 °C; the potential virus-free buds are isolated and grafted onto a virus-free (seedling) rootstock. This method usually results in a relatively high percentage of virus-free plants (Fridlund, 1980).

Virus-free grape-vines can be obtained by treating shoots (not meristems) in vitro for 21 days at 35 °C (Galzy, 1969a, b); plants in vivo could not tolerate these conditions. Walkey and Cooper (1976), showed that cucumber mosaic virus could be eliminated from in vitro grown shoot tips of *Stellaria media* by heat treatment.

19.2.2. *Meristem culture*

19.2.2.1. *History.* To get an idea as to how meristem culture was first used as a means of obtaining virus-free plants a short overview of the history is essential. White (1934a) showed that tobacco mosaic virus (TMV) was unequally distributed throughout the different zones of the tobacco root; a reduced concentration towards the top, with the root tip itself being free of the virus. Limasset and Cornuet (1949) proposed that there would also be such a virus gradient in the shoots of a tobacco plant, with the meristem being virus-free. Later it was conclusively shown that the root tips and shoot meristems were not always virus-free. In 1952 Morel and Martin had the brilliant idea of isolating in vitro the apical meristems of dahlias infected with the virus, from which they could obtain virus-free plants. They were the first to obtain virus-free dahlias and potatoes with the help of meristem culture (Fig. 19.2).

After Morel's discovery many publications pointed to it's importance. Although meristem culture is especially applied in horticulture, there are also some agricultural crops for which it is important: potatoes, *Trifolium, Lolium*, tobacco (Conger, 1981). Virus-free plants of some woody species have also been obtained using meristem culture: cherry, raspberry, currents, forsythia and grape.

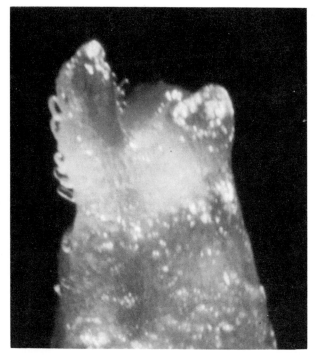

Fig. 19.2. Meristem of potato with two leaf primordia. Photograph from the Institute of Phytopathological Research (IPO), Wageningen.

As a result of these discoveries the question was naturally asked why the virus was unevenly distributed in the plant, and how this fact could be used with the help of meristem culture to obtain virus-free plants; whilst in other species (Quak, 1977) some virus particles were found in the meristems. Quak (1966) thinks there is competition in a meristem between on one hand production of virus particles and on the other cell production; in meristematic tissue during cell division the capacity for nucleic acid synthesis is being utilized for cell production to the detriment of virus multiplication; in cells below the meristem, which are increasing in size rather than dividing, virus multiplication takes place unheeded. Alternatively Quak (1966) proposed that the absence of vascular elements in the meristem greatly hinders the transport of virus particles. The possible absence of plasmodesmata in the meristem and the absence of vascular elements is for many people a very plausible explanation as to why the virus concentration is so low in meristems.

In the 50's it was thought (Quak, 1977) that the possible high concentrations of auxin and cytokinin in dividing (meristematic cells) would hinder the penetration of virus particles or that the virus particles would be inactivated by them; this theory has never been properly substantiated. Mellor and Stace-Smith (1969), proposed that enzymes were needed for virus multiplication which were absent in meristematic tissue; this also has to be substantiated. Martin-Tanguy et al. (1976), proposed that the low virus concentration in meristems is due to the presence of naturally occurring inhibitors; these inhibitors would be the reason why as a result of sexual reproduction seeds often remain free of viruses; this hypothessis has also still to be proven. The definite explanation of the production of virus-free plants by meristem culture is therefore still not yet possible.

19.2.2.2. *Accomplishment.* It is advisable to begin with growing shoots wherever possible when carrying out meristem culture (Fig. 19.3); when the meristem to be isolated is active (consisting of a meristematic and a subapical zone which enlarge quickly), the chance of eliminating the virus is greater.

PRODUCTION OF VIRUS-FREE PLANTS

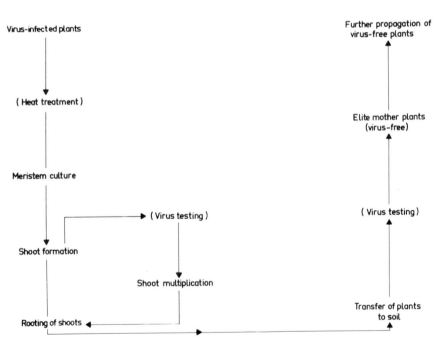

Fig. 19.3. Schematic representation of the various steps during the production of virus-free plants.

Shoots should be previously well cleaned. Any shoot leaves are removed where possible, and then the shoots (or buds if no leaves are removed) are dipped for a short time in 70% alcohol to remove trapped air. Following this, sterilization is carried out in diluted bleach or $Ca(OCl)_2$ and finally rinsed in sterile water. A few leaves are then removed under the stereomicroscope (magnification $20-40 \times$). A further sterilization process is then carried out using lower concentrations of bleach or having a shorter sterilizing time and then rinsing is carried out again with sterile water. In some laboratories 70% (v/v) alcohol is used for the second sterilization and no further rinsing takes place; if water is used for rinsing then it is possible that water drops can make isolation more difficult (reflection of light, poor visibility). The remaining leaves and leaf primordia are then removed one by one under a stereomicroscope; needles or pieces of razor blade mounted on an inoculation needle holder are often used for this part of the procedure. The shoot tip is held firmly in one hand (forceps or clean fingers) while the procedure is carried out with the other hand. The needle should be regularly sterilized during the procedure. When the meristem is practically free, it together with 1–2 leaf primordia (usually a cube of material) is cut out using a razor blade, and immediately inoculated (to prevent drying out) on a nutrient medium. To prevent drying out of the meristem a cold light source should be used for the binocular microscope used for the preparation. The meristem with leaf primordia is very small (0.1 mm in diameter; 0.2–0.4 mm in length).

There is a greater chance of obtaining virus-free plants if only the meristem is isolated; the chance of a meristem surviving without leaf primordia is very small as is shown by carnation (Morel, 1964; Stone, 1963). Accordingly isolation of larger meristems (with more leaf primordia) makes the chance of obtaining virus-free plants very small (Kartha and Gamborg, 1975).

Although in principle all shoot meristems of a plant are suitable as starting material, it appears that the chance of success is dependent on the type of bud or shoot (terminal or axillary) and/or the position (basal or terminal) of the bud (Styer and Chin, 1983).

The percentage of virus-free plants can also be very dependent on the season, as was shown by van Os (1964) in the case of carnation. To be sure of obtaining good root formation on the shoots it is sometimes recommended to isolate the meristems in a particular season; shoots of potato (Mellor and Stace-Smith, 1969) as well as those of carnation (Stone, 1963), root best when the meristem is isolated in the spring.

The composition of the nutrient medium is far from simple, since a meristem is very small to sustain in growth. Each different plant species, and sometimes even different cultivars of a single species require a special medium. The isolation medium is usually not the same as the rooting medium. Since deciding on a suitable medium (if there is no literature available) requires a great deal of work, meristem culture should only be used when a clone is completely infected and when heat treatment does not have the desired effect. It is strongly recommended that when choosing a nutrient medium the experience of others, found in the literature is used (Pierik, 1979: Bhojwani et al., 1986).

Meristems are usually isolated on solid media, although occasionally liquid media may be used (utilizing paper bridges). The pH usually lies between 5.4 and 6.0, and saccharose is the usual sugar (2–5% w/v). The following vitamins are often used: vitamin B_1, pyridoxin, panthothenic acid and nicotinic acid (Styer and Chin, 1983). Regulators are generally used at low concentration (0.1–0.5 mg l^{-1}); auxin may be necessary for root formation, auxin and cytokinin for stimulating cell division; GA_3 is sometimes added to obtain shoot extension (Morel, 1964). Sometimes as with *Pelargonium* cytokinin is added later and not added directly after isolation when it causes the formation of brown colouration (Debergh and Maene, 1977).

The normal growth temperature is 21–25 °C; most bulbous plants requiring a lower temperature. Higher temperatures (35–38 °C) are only used (usually alternate high and low temperatrue) to inactivate viruses (Walkey, 1980). Meristems are generally grown in fluorescent light (daylength of 14–16 h, irradiance about 8–12 W m^{-2}); sometimes fluorescent light is supplemented with a little red light (Styer and Chin, 1983). Sometimes a lower irradiance has to be used for the first few days after isolation (Debergh and Maene, 1977).

Sometimes the problem arises that a meristem produces a nice shoot, which will not root. Morel (1964), solved this problem in dahlias by grafting the virus-free shoot obtained in vitro on a virus-free seedling. When the plant is older it is possible to take cuttings without any problems. If only one virus-free shoot is available then it is possible to stimulate the development of lateral buds, so that more virus-free shoots can be obtained.

The percentage of isolated meristems that develop into virus-free plants varies, but is generally very small. The low success percentage is a consequence of loss due to infection, damage, drying out and browning; death or the initial failure to take can also be blamed on the incorrect nutrient medium, and problems with dormancy phenomena. If plants are eventually obtained then it is still necessary to check that they are virus-free. One virus-free plant is only enough to produce a virus-free clone if more virus-free plants can be vegetatively propagated quickly.

176

The mass production of virus-free plants is entirely dependent on the existence of methods for vegetative propagation.

19.2.3. *Heat treatment and meristem culture*

To increase the chances of obtaining virus-free plants in difficult cases (especially when more than one virus is present) heat treatment (Fig. 19.3) is often given at the beginning of the meristem culture; by this means the virus concentration is lowered and/or the virus-free zone enlarged. The length of the heat treatment (35-38 °C), varies from 5–10 weeks (Quak, 1977). The above procedure has been used successfully with potatoes (Fig. 19.4), chrysanthemum, carnation, and strawberry. For example Morel (1964), advised the storage of potato tubers at 37–38 °C for 1 month before starting meristem culture.

19.2.4. *Adventitious shoot formation, eventually followed by meristem culture*

In 1962 Brierley (see Morel, 1964) described the formation of virus-free adventitious bulbs in vivo on scales of *Lilium longiflorum*. Hildebrandt (1971) came to the same conclusion with gladioli. The regeneration of embryos from nucellus tissue of *Citrus* also usually resulted in the production of virus-free plants (Button and Bornman, 1971a, b and c; Bitters et al., 1972). Walkey et al. (1974) showed that flower-bud meristems of cauliflower were able to regenerate vegetative virus-free plants in vitro.

Fig. 19.4. A plantlet, produced in a test tube from a potato meristem is then transferred to soil. Photograph from the Institute of Phytopathological Research (IPO), Wageningen.

Mori et al. (1982) showed that adventitious shoots which were obtained in vitro from TMV infected leaf explants of tobacco were virus-free; the same was shown with the development of adventitious shoots for grape (Barlass et al., 1982).

Obtaining virus-free plants by the adventitious shoot method in vitro has had great success with lily (Allen, 1974; Asjes et al., 1974), and hyacinth. For this reason, virus infected lily scale explants are allowed to regenerate adventitious bulblets in vitro; when the meristems are about 1 mm in size they are transferred to another nutrient medium for further development. Hakkaart et al. (1983), proposed that the adventitious-shoot method combined with meristem culture produced plants in *Kalanchoë* 'Yucatan', which were free from any virus symptoms.

The adventitious-shoot method is used in a different way to obtain virus-free plants with some other plant species (petunia, tobacco, cabbage). With these plants it appears that particular areas of the leaf are virus-free, while the plant as a whole is infected with the virus. Isolation of virus-free zones and regeneration of adventitious shoots resulted in the production of virus-free plants (Murakishi and Carlson, 1979).

If the adventitious shoot method is used for chimeric plants such as the variegated *Pelargonium*, then the variegated form is lost, as adventitious shoot formation usually takes place from one cell (layer) only (Cassells et al., 1980).

19.2.5. *Virus-free plants produced from callus and protoplasts*

In 1962 Cooper (see Hildebrandt, 1977) showed that after a few subcultures tobacco callus can become virus-free; apparently callus tissues can escape virus infection; young meristematic active cells appear especially more resistant to TMV than older inactive cells. The results of Cooper's research were later corroborated by Chandra and Hildebrandt (1967). Abo-El-Nil and Hildebrandt (1971) isolated anthers of *Pelargonium* and obtained from these callus and plantlets which appeared to be virus-free; the same was shown with potato callus that was infected with potato X virus. Walkey (1978) claimed that callus of *Nicotiana rustica* could be made virus-free by heat treatment.

Mori et al. (1982) isolated protoplasts from tobacco leaves which were infected with TMV; from these protoplasts they regenerated virus-free plants. Isolation of protoplasts from the dark green zones (with few if any virus particles) of tobacco leaves infected by TMV also produced virus-free plants.

It is however not realistic to assume that the production of virus-free

178

plants via callus and protoplasts will have practical importance, since with this type of culture mutations are very common.

19.2.6. *Grafting of meristems on virus-free (seedling) rootstocks (micro-grafting)*

If it is not possible to induce a meristem to grow, or a shoot obtained from meristem culture will not form roots, then it is possible to graft the meristem onto a virus-free (seedling) rootstock, which is grown or propagated in vitro. Micro-grafting is extremely important with woody species since with this group meristem culture is often impossible. The concept of micro-grafting has arisen from the work of Morel and Martin (1952) with dahlias. The first successful micro-grafting was carried out by Murashige et al. (1972) and Navarro et al. (1975), working with *Citrus*, who eliminated two viruses. A review of the micro-grafting of *Citrus* is given by Litz et al. (1985). Later other plants were used such as apricot, and the wine grape (Anonymus, 1978a), *Eucalyptus* (Damiano et al., 1986), *Camellia japonica* (Creze, 1985), peach (Mosella, 1979; Mosella et al., 1979), apple (Huang and Millikan, 1980). With apricot and the wine grape the micro-grafting was preceded by a heat treatment. Jonard et al. (1983), and Kartha (1986) have given an appraisal of the use of micrografting in fruit crops. For an extensive analysis of the (in)compatibility phenomena which arise during in vitro micro-grafting, the reader is referred to a review article by Jonard (1986).

19.2.7. *Virus identification*

It cannot be visually determined that a plant is virus-free (free of the known viruses), since there are many latent virus infections. The potential 'virus-free' plant must therefore be assayed, which can be carried out as follows (Hakkaart et al., 1980; Reuther and Sonneborn, 1981; Maat, 1980; Maat and Schuring, 1981):

1. Test plants. The sap from one of the plants to be assayed is smeared onto the leaf of a test plant, which has already been treated with carborundum powder. If virus is present in the sap then after a few days the test plant begins to show characteristic symptoms. Important test plants are: *Chenopodium amaranticolor, Chenopodium quinoa, Gomphrena globosa* and many different species of tobacco. Disadvantages of this type of assay are the length of time needed and the requirement for extensive growth facilities.

2. Electron microscopy. Little use is made of this expensive method in practice, since few laboratories possess the specialized equipment and trained personnel to carry out the procedures.
3. Serology. If a rabbit is inoculated with proteinaceious receptors, such as a plant virus, then specific antibodies are formed. If a drop of centrifuged plant sap is added to a drop of anti-serum from the blood of the rabbit, then if the virus is present precipitation will take place. On the basis of this phenomena many, sometimes extremely sensitive serological methods have been developed. One of these is the ELISA (enzyme-linked immunosorbent assay). Details of the use of serological methods can be found in Kartha (1986) and Long and Cassells (1986).

19.3. Production of bacteria- and fungal-free plants by meristem culture

It appears that bacteria- and fungal-free plants are also obtainable by the use of meristem culture (LaMotte and Lersten, 1972; Tramier, 1965). The most important genera of bacteria to be eliminated are: *Erwinia, Pseudomonas, Xanthomonas* and *Bacillus*; and the most important gen-

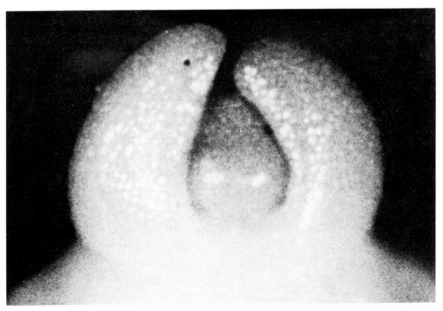

Fig. 19.5. Apical meristem of carnation with 2 leaf primordia, which is isolated in vitro to obtain disease-free plants. Photograph from the Institute of Phytopathological Research (IPO), Wageningen.

era of fungi: *Fusarium, Verticillium, Phialophora* and *Rhizoctonia*. A rich nutrient medium (in which peptone, tryptone or yeast extract are added) is sometimes used to quickly determine whetehr a plant is bacteria- or fungal-free.

The first research (Baker and Philips, 1962) on the production of fungal-free plants by meristem culture concentrated on the carnation (Fig. 19.5), from which *Fusarium roseum f. cerialis* was eliminated. Gladioli are often attacked by *Fusarium oxysporum f. gladioli,* which is very difficult to combat. As a result of fungal attack in France, already many gladioli cultivars have been almost completely lost. Tramier (1965), successfully obtained non-infected plants by meristem culture, which were then propagated in surroundings where the fungus was not found.

In 1977 Rudelle showed that for carnation two fungi (*Fusarium oxysporum f. dianthi* and *Phialophora cinerescens*), and two bacteria species (*Pseudomonas carophylli* and *Pectobacterium parthenii*) could be eliminated with the use of meristem culture. Theiler (1981) obtained bacteria-free plants of *Pelargonium* which had been initially infected with *Xanthomonas pelargonii*. Meristem culture has also been used successfully in the group of *Begonia elatior* hybrids to eliminate the bacterium *Xanthomonas begoniae* (Hannings and Langhans, 1974; Reuther, 1978; Moncousin, 1979; Theiler, 1981), and with strawberry to eliminate *Phytophthora fragariae* (Seemüller and Merckle, 1984).

20. Vegetative propagation

20.1. General introduction

In principle plants can be propagated in two ways: vegetatively (asexual, also called cloning), and generatively (sexually, by seeds). Both types of propagation may be impossible under certain conditions. When generative propagation is unsatisfactory (no seeds are formed, too few seeds are formed, the seeds quickly loose their germination viability) then vegetative propagation is often adopted. Generative propagation is equally unsatisfactory when a (very) heterogeneous progeny is obtained, due to its strong heterozygosity.

In vivo vegetative propagation (via cuttings, splitting or division, layering, earthing up, grafting and budding) has for many years played an important role in agriculture; e.g. with potatoes, apples, pears, many ornamental bulbs and tuberous plants, many arboricultural crops, carnations, chrysanthemums, etc. Vegetative propagation is also very important in plant breeding: parent lines have to be maintained and propagated vegetatively for seed production; cloning is often required for setting up gene banks; adventitious shoot formation is needed to obtain solid mutants after mutation induction.

The classical methods of in vivo vegetative propagation often fall short of that required (too slow, too difficult, or too expensive) or are completely impossible. In the last 10 years, since the discovery that plants can be more rapidly cloned in vitro than in vivo, knowledge concerning in vitro vegetative propagation has grown quickly; this holds equally true for plants from temperate, subtropical as well as tropical regions. It has now become possible to clone species by in vitro culture techniques that are impossible to clone in vivo.

This Chapter gives an outline of the different methods of in vitro vegetative propagation: single-node cuttings, axillary branching, regeneration of adventitious organs (roots or shoots) on explants (Fig. 20.1),

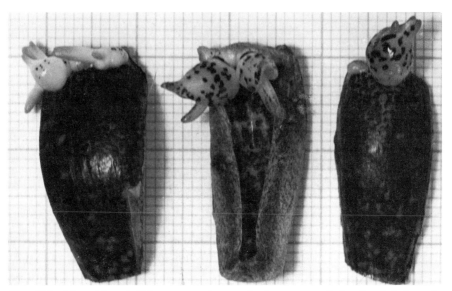

Fig. 20.1. Adventitious bulblet regeneration at the bases of apolarly placed bulb scale explants of hyacinth. By applying this method vegetative propagation can be accelerated (Pierik and Woets, 1971). Photograph taken 12 weeks after isolation.

formation of adventitious organs and somatic embryogenesis on callus tissue and regeneration of plants from cell suspensions and protoplasts. Some particular forms of vegetative propagation such as the cloning of orchids, and micro-grafting are not dealt with in this Chapter since they have already been fully covered in other Chapters.

To propagate plants in vitro by adventitious organ or embryo formation, in principle, it is necessary that they are capable of regeneration. The ability to regenerate is determined by the genotype, the environmental conditions (nutrient supply, regulators and physical conditions), and the developmental stage of the plant. It is well known that some families and genera have a high regeneration ability: *Solanacea (Solanum, Nicotiana, Petunia, Datura* and *Lycopersicon), Cruciferae (Lunaria, Brassica, Arabidopsis), Gesneriaceae (Achimenes, Saintpaulia, Streptocarpus), Compositae (Chichorium, Lactuca, Chrysanthemum), Liliaceae (Lilium, Haworthia, Allium, Ornithogalum).* Chapters 6 and 13 pointed out the importance of the environmental factors in enabling the plant to regenerate to its maximum capability. The developmental stage of the starting material (Chapter 12) is the next most important factor: juvenile plants have a greater regeneration capacity than adult plants. Since adult plants are generally used for vegetative propagation, this means that especially in the case of woody species, an attempt should be made to rejuvenate them before use. Rejuvenation can sometimes be achieved by: meristem culture; the cutting from a cutting method (also used in vivo); grafting a meristem or shoot tip on a (seedling) root stock; isolating from zones which are still juvenile; pruning severely to stimulate the production of lateral juvenile shoots. Although adventitious shoot- and embryo-formation often result in rejuvenation, they are not as satisfactory since mutations are more likely. Rejuvenation by means of meristem culture, despite the difficulties associated with this method, is still the most favoured technique since it: maintains genetic stability, eliminates fungi and bacteria, and can sometimes result in the additional advantage of obtaining virus-free material.

If problems are encountered when using adult plant material for vegetative propagation in vitro, then a model study should be set up using seedling (juvenile) material (Fig. 20.2). Seeds are sterilized and sown in vitro; and then the seedlings obtained (often genetically heterogeneous) are used for in vitro organ formation (by axillary or adventitious shoot formation). Using all the information obtained in this way it is then possible to try in vitro propagation using material from selected adult plants. This has been successfully carried out with *Anthurium andreanum* (Pierik, 1976a).

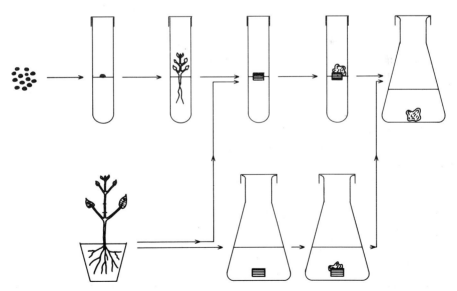

Fig. 20.2. Juvenile starting material (seeds) for the initiation of callus culture. In the upper left part a sterilized seed is isolated in vitro and a seedling arises. The seedling can also be obtained by sowing seeds in a pot (bottom left). Subsequently explants are isolated on a solid medium (top) or in a liquid medium (bottom). When callus induction has taken place, the callus is subcultured in an Erlenmeyer flask (upper row, right).

One of the most striking differences between herbaceous and woody species, is that the latter are far more difficult to clone in vitro. The reasons for this are (Kunneman-Kooij, 1984; Pierik, 1975; Bonga and Durzan, 1982):

1. Woody species have a relatively weak regenerative capacity when compared to herbaceous species.
2. Research with trees and shrubs was initiated later.
3. The induction of rejuvenation is generally extremely difficult in woody species.
4. The multiplication rate is much lower with woody species than herbaceous species.
5. Dormancy plays a role in the case of trees and shrubs; buds do not open and stem elongation fails to take place.
6. Topophysis plays a greater role with woody species.
7. Woody species are more liable to be effected by excretion of toxic substances into the nutrient media.
8. Sterilization is more difficult with woody species, which are usually grown outside.
9. Trees and shrubs can often only be selected for cloning when they are adult. Since adult material is often very difficult if not impossible to propagate in vitro, insurmountable problems are often encountered.
10. Genetic variation in trees is generally greater than in agricultural and horticultural crops, giving rise to variable results.
11. Greenhouse material is not usually available for woody species, and explants must be taken from field grown trees. There will be considerable variation in the explants due to the different growth conditions and annual fluctuations in the climate.

The widespread use of in vitro cloning is indicative that it has many advantages. These are listed below (Pierik, 1975; Anonymous, 1980a; van Assche, 1983; Gebhard et al., 1983; Kunneman-Kooij, 1984):

1. Propagation in vitro is more rapid than in vivo.
2. It is possible to propagate some species by in vitro culture which cannot be propagated in vivo; this is sometimes due to rejuvenation which is only possible in vitro.
3. The growth of in vitro propagated plants is often stronger than those cloned in vivo; this is mainly due to rejuvenation and/or the fact that they are disease-free. Cameron et al. (1986) demonstrated that the increased vigour and productivity of micropropagated strawberry plants appeared to be related to their gas exchange properties.

4. Using in vitro culture is possible, in principle, to have disease-free propagation, either by strong selection of the starting material or by first making the starting material disease-free. Disease-free transport of plants in vitro (without soil) is also made possible.

5. Since relatively little starting material is needed for in vitro culture, vegetative propagation can be started after strong selection has taken place. In this way field trials can be quickly established, enabling further selection to take place.

6. In vitro propagation can produce great savings in fuel costs, greenhouse space, etc. The area needed for raising stock plants, and propagating beds is decreased by in vitro culture.

7. Due to the perfect conditioning (nutrient media and physical conditions), enabling precise timing of cutting production, the effect of the seasons can be eliminated and year round production achieved.

8. The propagation in vitro on its own root system makes grafting and budding onto root stocks unnecessary, saving a great deal of work. This will be of importance especially for species such as rhododendron, rose, and lilac.

9. For the professional plant breeder there are also the following additional advantages:
   — A new cultivar can be made commercially available more quickly with in vitro than by in vivo propagation.
   — With in vitro culture it is often possible to build up (small) clones more quickly, which can be used as parents for the production of F_1-hybrids.
   — Plant breeders can readily obtain solid mutants during the induction of adventitious shoots.
   — In vitro culture is especially useful for setting up gene banks (stored disease-free at low temperatures).
   — In relation to genetic manipulation (somatic hybridization etc.) regeneration of plants from protoplasts and cells is indispensable.
   — Some plants need to be maintained and propagated vegetatively: sexually sterile plants (haploids, sterile mutants, male sterile lines for hybrid production), rare aneuploids or plants with unusual combinations of chromosomes that may be lost by propagating through seeds, and specific heterozygous gene combinations (Dale and Webb, 1985).

It is also beyond doubt that in some cases in vitro cloning can have disadvantages:

1. In some in vitro propagation systems, the genetic stability is weak. This is discussed in more detail in Chapter 21.
2. In vitro produced plants may display troublesome effects in vivo: bushiness (the repeated production of side shoots), complete reversal to the juvenile phase.
3. Especially with woody plants the induction of rooting is often extremely difficult. In the case of other plants the roots formed in vitro may not be functional, and have to be replaced in vivo by new roots adapted to the soil.
4. The transfer of some plants from the test tube to soil is very difficult to achieve. This was discussed fully in Chapter 14.
5. When plants, which are to be cultivated outside, are mass-cloned in vitro and then transplanted to soil, the danger arises that the clone will be killed by pathogens as soon as a new strain of a micro-organism arises to which the plant is not immune. In the case of tropical plants such as the oil palm, a so-called multi-clonal plantation has been considered to avoid this problem. In protected cultivation intensive pest control can be applied so that this cannot occur.
6. Regenerative capacity can be lost by repeated culture of callus and cell suspensions.
7. In some cases sterile isolation is extremely difficult to realize.
8. In vitro cloning is very labour intensive, which results in a relatively high price for the plants produced (Smith, 1986). The use of special techniques (fermentation culture), which will save labour costs are still at an experimental stage (Takayama et al., 1986).

After listing the advantages and disadvantages of in vitro cloning, perhaps we should itemize the criteria desired for its success:

1. Gene stability (no mutations).
2. Strong selection for disease-free starting material, followed by disease-free propagation.
3. The transfer from test tube to soil should not be too difficult.
4. Regeneration ability should not be lost.
5. The in vitro propagation should not be too complicated otherwise it will be rejected.
6. It should be economically viable.

It is difficult to estimate the economic viability of commercial in vitro propagation. Until now, de Fossard (1977), Jacob et al. (1980), van Assche (1983), Kyte (1983), and Sluis and Walker (1985), are the first people to have given some thought to this problem. The development of large commercial tissue culture laboratories in Western Europe and the U.S.A. indicate that the commercial and economic problems are not insurmountable.

Before discussing the different ways it is possible to vegetatively propagate in vitro, it is necessary to explain the different phases that are recognized (Murashige, 1974; Debergh and Maene, 1981):

Phase 0: This covers what occurs before in vitro culture begins: the correct pre-treatment of the starting material, as far as possible keeping the plants disease-free (no insects in the greenhouses, only water in the pots, keep the plants relatively dry, good disease prevention etc.).

Phase 1: This covers the sterile isolation of a meristem, shoot tip, explant, etc. If there is internal infection special techniques are needed (see Chapters 9 and 19). In this phase there is only one important requirement: to accomplish non-contaminated growth and development.

Fig. 20.3. Shoot propagation (Phase 2) of *Nidularium fulgens* in liquid medium. Shoot propagation takes place by axillary branching (Pierik and Steegmans, 1984). Photograph taken 12 weeks after isolation.

Phase 2: The propagation phase. The primary goal in this phase is to achieve propagation without loosing the genetic stability. Propagation can be carried out in a number of ways; these are discussed in detail later in this Chapter (Fig. 20.3).

Phase 3: This involves the preparation of the shoots and plants, obtained in phase 2, for transfer into soil. This can involve: stopping the axillary shoot formation, and initiating shoot elongation. Subsequently root formation must be induced, either in vitro or later in vivo.

Phase 4: This covers the transfer from test tube to soil, and the establishment of the plantlets. More information is available in Chapter 14.

The volume of literature concerning in vitro vegetative propagation has become so large that it is beyond the scope of this book to discuss it extensively, or to consider all the different species so far studied. For this reason, only the general principles will be set out; for more detailed information about a particular species the bibliography by Pierik (1979) and the handbooks by George and Sherrington (1985), Bhojwani et al. (1986) and Bajaj (1986, 1986a) should be consulted. More detail is given about propagation systems that are popular in practice and found to be satisfactory, rather than systems that have been found to be unsatisfactory (such as callus-, suspension- and single-cell cultures). The problem of genetic (in)stability is discussed separately in Chapter 21. The application of micro-grafting in vitro has already been considered in Section 19.2.6.

20.2. Single-node culture

Here we mean the isolation from a bud, together with a piece of stem, with the purpose of forming a shoot by allowing the bud to develop. This method is the most natural method of vegetatively propagating plants in vitro, since it is also applicable in vivo. Each bud that is found in the axil of a leaf, just the same as the stem tip, can be isolated on a nutrient medium; an attempt is then made to allow this bud to develop in vitro. The buds in the axils of the newly formed leaves can then also be subcultured, and allowed to develop, etc. When enough shoots have been obtained, these must be rooted, and then ultimately transferred to soil. This isolation of buds and shoot tips (Fig. 20.4), is, in principle, a technique where no cytokinin is added to prevent apical dominance. The following points should be borne in mind when applying the method:

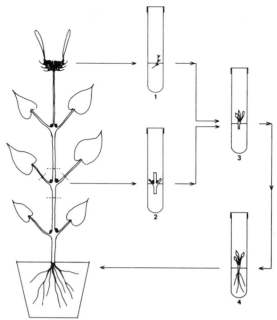

Fig. 20.4. Schematic illustration of vegetative propagation by the single-node technique. In the axils of the leaves and the involucral bracts of the inflorescence vestigial nodes are found which, after isolation (1,2) give rise to a shoot (3) which in 4 has regenerated roots. Additional nodes can be isolated by allowing the shoot in 3 to elongate.

1. Isolation by this method is practically impossible when dealing with rosette plants (such as *Bromeliaceae,* gerbera, etc.), and when the chances of infection are high. The axillary shoot method (Section 20.3) is used for rosette plants.
2. To reduce the chances of infection it is best to isolate closed buds (Anonymous, 1978a). If there are internal infections, then this method is unusable and it is first necessary to resort to meristem culture.
3. The rate of propagation is dependant on the number of buds available; if this number is small then the propagation is very slow. The more leaves that are initiated with time, the more buds become available, and the more rapid is the propagation.
4. Dormancy, especially of woody plants from temperate climates often presents problems (Anonymous 1978a). To break the dormancy a good understanding of the problems associated with dormancy (breaking) is needed. As was pointed out in Chapter 12, the starting material itself plays an important role in this respect (position, age, season of the year at the time of isolation, etc.). Physical factors (Chapter 13) are also of great importance as is the composition of the

nutrient medium (Chapter 6). Ways of breaking dormancy after in vitro isolation, especially for woody plants are: low temperature treatment (0–5 °C), long day treatment (16 h light per day), and low temperature, followed by long days. In some cases gibberellins and/or cytokinins can promote dormancy breaking of the buds. In many cases (especially with herbaceous plants) neither cold nor long days are required to break the dormancy. Etiolation (due to darkness or low irradiance) may result in dormancy breaking of buds.

5. Juvenile shoots, especially of woody plants, grow much more quickly than adult shoots, which therefore need to be rejuvenated.

6. The rooting of adult shoots, certainly in the case of woody plants is far from easy, and initial rejuvenation is also recommended. For more information on the induction of adventitious root formation Section 20.4.2 should be consulted.

The first research into isolation of buds, and the subsequent rooting of the shoots formed, was carried out on asparagus (Galston, 1947, 1948; Gorter, 1965; Andreassen and Elison, 1967). Here the rooting of the shoots was particularly difficult, and it has been shown that darkness, as well as a high NAA concentration, is needed. It is also recommended that the rooting takes place on a liquid medium (better O_2 supply); to induce root elongation the NAA concentration should be lowered. In addition with the asparagus the induction of a root stock is also necessary, which is far from easy.

A very special sort of single-node cutting is the transformation of isolated flower bud primordia (originating from parts of the inflorescence) in vegetative shoots; the classic example of this is cauliflower (Crisp and Walkey, 1974).

The single-node method (Fig. 20.5) has had a good measure of success with many different plants: potato (Morel and Martin, 1955), *Manihot esculenta* (Kartha et al., 1974), *Vitis rupestris* (Galzy, 1969b), pear (Quoirin, 1974), rose (Elliott, 1970), *Hedera helix* (Hackett, 1970), *Salix babylonica* (Letouzé, 1974), etc.

Recently the single-node method has been applied to *Eucalyptus grandis* and *Araucaria cunninghammii* (Mott, 1981), and in the latter case with pronounced orthotropy/plagiotropy it can only be used for orthotropically (upright) growing shoots. With coffee, on the other hand (Staritsky, personal communication), there is spontaneous transition from plagiotropic to orthotropic growth.

From our own research (Pierik and van der Kraan, unpublished data), it has been shown that the single-node method is also of use with plants such as tomato, cucumber and aubergine: sterile isolated seeds of these

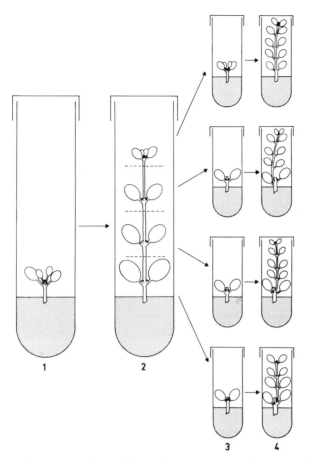

Fig. 20.5. Schematic representation of the single-node method of vegetatively propagating plants.

plants form a plant in vitro, from which the shoot tip as well as all the single-nodes can be used. The advantage of this method with plants such as tomato (with a very strong apical dominance, which cannot be propagated by the axillary shoot method) is that they can be cloned very efficienty in vitro. Repeated single-node culture was also successfully applied to *Syringa vulgaris* (Pierik et al., 1986): stem elongation in this species is obtained efficiently by the application of zeatin(riboside) or 2-iP, and by low irradiance.

20.3. The axillary bud method

In principle, this method is very similar to the single-node method: the biggest difference being that with the latter method elongated plants (with stem) are used almost exclusively, and cytokinin is not usually required for the buds to develop.

With the axillary bud method (Fig. 20.7) a shoot tip is isolated, from which the axillary buds in the axils of the leaves develop, under the influence of a relatively high cytokinin concentration. This high cytokinin concentration stops the apical dominance and allows axillary buds to develop. It is interesting to note that the principles of this method were already known in 1925: the apical meristem restricted the growth of the axillary buds; by removal of the apical meristem, the apical dominance is also lost. After the first cytokinin (kinetin) was discovered it was realized that the apical dominance could also be broken by the addition of cytokinin. If a shoot tip has formed a number of axillary- or side-shoots, then they can be inoculated onto fresh media containing cytokinin; new axillary shoots form etc. When enough shoots have been obtained they can be rooted and the resulting plantlets planted out into soil.

In practice, the single-node method is often used in combination with the axillary bud method (Fig. 20.7); an eye is allowed to develop, and then cytokinin is used to induce axillary shoot formation.

Internal infections are a problem when using the axillary shoot method. Especially in the case of rosette plants, sterile isolation is sometimes difficult. For this reason shoot tips are isolated, which are as small as possible, to give the best chance of obtaining a sterile culture. However, it must be born in mind that the smaller the isolated the shoot tip the less likely is it that it will become established. When internal infection is present then meristem culture should be applied first. In all cases it is advisable to try a rich nutrient medium (with peptone or tryptone), to test is the shoot-tip is free of internal infections. To be more certain of having sterile cultures a shoot tip should be cut lengthwise and inoculated flat on a rich nutrient medium; internal infections are then more rapidly expressed. Cutting the tip lengthwise can result in the death of the plant, so this check should not be carried out if there is a limited amount of starting material available. Finally it should be pointed out that the test on rich nutrient medium is by no means certain in all cases.

Despite the problems of isolating sterile shoot tips, the axillary shoot method, certainly for rosette plants, has become in practice far the most important propagation method in vitro. The reasons for this are as follows:

194

1. The method is generally simpler than other methods of propagation.
2. The rate of propagation is relatively fast.
3. Genetic stability is usually preserved (for exceptions see Chapter 21).
4. The growth of the resulting plants is very good, perhaps due to the rejuvenation and/or the lack of infections.

In the literature it is frequently found, especially in the case of meristems, but also shoot tips of adult plants that they are rejuvenated after a number of subcultures (in stages); this rejuvenation is more or less the same as produced by the 'cutting of a cutting' method which is used in vivo to produce rejuvenation (partially). The best example of rejuvenation after meristem culture in vitro can be found in the orchids (Section 18.2). There are strong indications (Margara, 1982; Moncousin, 1982), that the degree of rejuvenation is dependant on the size of the isolated meristem, the number of subsequent subcultures, and the concentration of regulators in the medium. Similarly the rate of propagation with the axillary shoot method is definitely not constant, but dependent on the number of subcultures; the more subcultures that have taken place, the quicker is the rejuvenation. Successful rejuvenation of 30 year old *Eucalyptus grandis* trees was achieved after 12 repeated subcultures during a period of 1 year (Brutti and Caso, 1986). Similar results were obtained with 30–50 year old lilac *(Syringa vulgaris)* shrubs by Pierik et al. (1986).

The change from a adult to a juvenile phase is shown as follows (Moncousin, 1982):

1. Morphological characteristics (e.g. leaf form) change.
2. The rooting potential increases.
3. The peroxidase activity of the shoot tip changes.
4. The K/Ca ratio in juvenile *Sequoia sempervirens* appears to be higher than in adult material; if the plant material is rejuvenated with a number of subcultures then the K/Ca ratio increases (Boulay, 1986).
5. The endogenous IAA/ABA ratio is higher in juvenile material of *Sequioa sempervirens* than in adult material (Boulay, 1986; Fouret et al., 1986).

Since it is not possible to give a complete list of all the factors that influence axillary shoot formation, a few of the most important points are dealt with below:

1. The cytokinin requirement is extremely variable (sort of cytokinin and concentration). For example seedlings of *Bromeliaceae* (Pierik, unpublished data) are able to form axillary shoots without any cytokinin in the medium. For many plants it holds true that, the more subcultures are made the cytokinin requirement lessens, as a result of habituation and rejuvenation. Most plants react very good to BA and often even less to kinetin or 2iP. In the case of the *Gymnospermae* BA is added, almost without exception to efficiently stop apical dominance (Kunneman-Kooij, 1984). With plants such as *Rhododendron* only 2iP is applied. The cytokinin concentration should be adjusted to account for the plant species and the particular cultivar in use. Kunneman-Kooij (1984) gave the concentrations of cytokinin suitable for a number of woody plants.

2. It also follows that it is very important to correlate the cytokinin concentration with the developmental stage of the material. Juvenile material requires less cytokinin than adult material. The developmental stage is also dependent on the number of subcultures that has taken place. For example with *Sequoia sempervirens* (David, 1982) it has been demonstrated that juvenile tips form better axillary shoots than adult plants.

3. It is often advisable to choose a low auxin concentration, together with a high cytokinin concentration; the cytokinin/auxin ratio usually being 10/1. In some cases, such as with the *Bromeliaceae,* auxin and cytokinin work synergistically (Fig. 20.6).

4. It is sometimes recommended, that the shoot tip should be allowed to develop before the cytokinin concentration is increased to induce axillary shoot formation.

5. If there is no satisfactory axillary shoot development then by either removing, or killing the apical meristem, the axillary branching can be improved. Pierik and Steegmans (1986) showed that with *Yucca flaccida* and *Yucca elephantipes,* the axillary shoot formation could be stimulated by longitudinally cutting the shoot tip.

6. Callus formation, as a result of too high a cytokinin concentration must be avoided; this can result in the formation of adventitious shoots, with the possibility of mutations occurring.

7. Axillary shoot formation can be promoted in some cases (e.g. *Bromeliaceae)* by using a liquid, rather than a solid medium. Whether the liquid medium should be shaken is dependent on the experimental plant.

8. It has been demonstrated with a wide range of species that thidiazu-

Fig. 20.6. Axillary shoot formation, in a liquid medium, with shoots of *Vriesea* 'Poelmanii'. The top and bottom rows received 0.08 mg l^{-1} BA and 0.08 mg l^{-1} BA plus 0.01 mg l^{-1} NAA (auxin) respectively. There was an almost complete lack of axillary branching in the absence of auxin. Photograph taken 16 weeks after isolation.

ron and several substituted pyridylphenylurea compounds stimulate axillary branching, more than the classical cytokinins (Read et al., 1986).

9. Changes in shoot proliferation can occur with repeated in vitro culture, although the reasons why are not known. Norton and Orton (1986) observed that shoot proliferation in the *Rosaceae* increased over the first few subcultures (1–3 depending on the species), and then gradually decreased at all BA concentrations (except in *Chaenomeles japonica,* in which the decrease only occurred at 5 mg l^{-1}); they also found that shoot length and leaf size decreased and callus formation increased after several subcultures.

The axillary shoot method was first used by Hackett and Anderson (1967), for carnation, then by Adams (1972) and Boxus (1973, 1974) for strawberry, and by Pierik et al. (1973, 1974, 1975a) and Murashige et al. (1974) for gerbera. It is striking that in recent years this method is rarely used for carnation, while it is used on a large scale for strawberry and gerbera. Further information concerning strawberry and gerbera, is found later in this Section.

The axillary shoot system (Fig. 20.7), developed for strawberry by Box-

197

us (1973, 1974) and Boxus et al. (1984), in Belgium, is known as micro-propagation via axillary buds. Boxus first carried out meristem culture, to obtain disease-free material, before adding cytokinin to induce axillary shoot formation. In practice not more than 12 propagation cycles are carried out with strawberry, since too many cycles can result in physiological deviations (Debergh, personal communication). In vitro propagation of strawberries appears to have the following advantages (Aerts, 1979):

1. Since meristem culture is used initially, and tests carried out for virus infection, virus-free clones are obtained, resulting in a considerable increase in production.

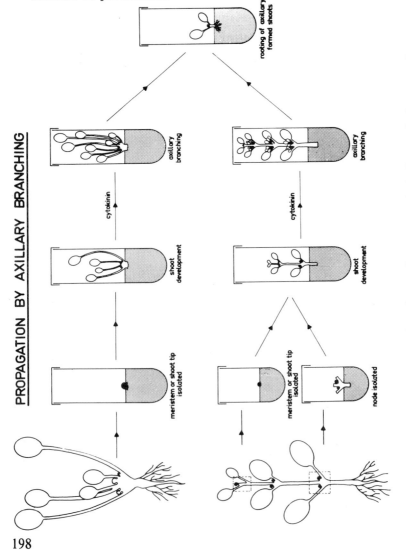

Fig. 20.7. Schematic representation of the axillary bud method of vegetatively propagating plants. Upper row: as applied to rosette plants. Bottom row: as applied to plants which elongate.

198

2. The production is also helped by the fact that the material is free from other pathogens (nematodes and soil fungi).
3. Propagation is rapid, this also being of importance when propagating new and disease-free cultivars.
4. Propagation is no longer dependent on the season, making timing and planning much more simple. Production peaks can be avoided, since cultivars can be stored at low temperatures.

Cytokinin also appears to break the dormancy of buds of gerbera. In 1973 Pierik et al. discovered that if a capitulum explant with involucral bracts, is isolated in vitro, then shoots can develop from the meristems, which are found in the axils of the involucral bracts. The development of shoots only occurred if a dark period of 4–6 weeks was given initially, together with a high cytokinin concentration. Shoots developed in the light (given after the dark period) could be rooted on another nutrient medium containing auxin, but without cytokinin. Further details are given in Pierik et al. (1974, 1975a). Murashige et al. (1974), discovered another system for gerbera. They isolated a shoot tip of gerbera, which had a number of leaf primordia; the axillary buds in the axils of the leaves, developed under the influence of a high cytokinin concentration. The newly developed shoots could be transferred to fresh media, on which axillary shoot formation could again take place. By this method it is possible to build up a clone of gerbera in a relatively short period of time. Murashige's system is now used worldwide on a large scale.

Strawberry also has the same problem as gerbera which was discovered later for other plants: bushiness. When axillary shoot formation is stopped (by leaving out cytokinin), the shoot tips continue to form side shoots (even during and after rooting). This results in the development of bushy plants, which have smaller flowers and/or fruits than normal. This results in higher picking costs in the case of strawberries, where fruits are picked individually. Pierik et al. (1982), showed that the cytokinin requirement for axillary shoot formation in gerbera is dependent on the cultivar used: it was also found that a high cytokinin concentration not only caused bushiness, but also gave rise to different leaf forms, callus formation and adventitious shoot formation. Adventituous shoot formation can result in mutations.

The axillary shoot method for in vitro propagation has now been applied to so many different plant species that it is impossible to review them all here. However it is worth mentioning that the axillary shoot method for the propagation of rootstocks and cultivars on their own rootstock has become very popular in the fruit production industry, especially in England and Italy. This method has the following advantages in fruit production:

1. Rapid propagation is possible of new rootstocks and cultivars and from disease-free plant material.
2. By propagation of cultivars on their own roots, grafting and budding onto rootstocks becomes unnecessary, saving labour costs.
3. In vitro propagation also cuts down wastage of material, which is found when unsuccessful grafting takes place during in vivo propagation. However, it should be borne in mind that wastage may also occur in vitro, due to unsuccessful transfer of the material to soil.
4. Propagation is independant of the seasons, provided that the material can be stored at low temperatures or planted out in a greenhouse.

It is surprising that propagation of fruit trees is hardly used in some countries, such as in The Netherlands; the reasons for this are as follows:

1. There is a good system for producing and making available virus-tested material through the use of grafting on rootstocks. This is not the case in England and Italy.
2. The in vitro method which is labour intensive is possible too expensive to introduce in The Netherlands, where wages are relatively high.
3. The Dutch fruit growing industry is afraid of introducing mutations. However, on the basis of literature research (Németh, 1986), this has not proved to be a problem for apple, pear, plum, cherry or peach.

Rooting of axillary shoots will be dealt with in Section 20.4.2.

20.4. Regeneration of explants

20.4.1. Introduction

The regeneration (the adventitious formation or new formation) of organs (Fig. 20.8), that were not present at the time of isolation is an extremely complicated process because:

1. Existing correlations have to be broken before new ones can be built up, which will lead to organ regeneration.
2. It consists of a number of processes:
   — De-differentiation of differentiated cells (possibly leading to rede-termination and rejuvenation of cells).

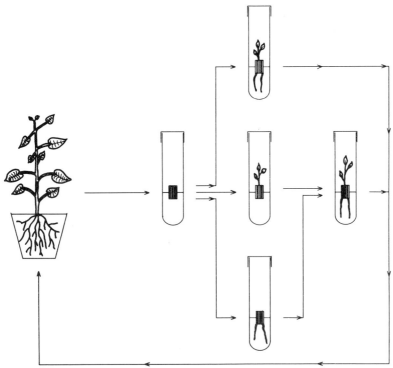

Fig. 20.8. Schematic illustration of vegetative propagation through the regeneration of adventitious roots and/or shoots. In the upper test tube simultaneous regeneration of a shoot and roots occurs. In the middle test tube only a shoot and in the bottom test tube only roots regenerate. If only a shoot or a root develops, the material is subcultured onto a medium of different composition so that a complete plant can be obtained (right).

    — Cell division, sometimes followed by callus formation; when cell division is directed organ initiation can begin.
    — Organ initiation (formation).
    — Organ development.
3. It is limited both qualitatively and quantitatively by a large number of factors: those inherent in the plant material, the growth conditions of the plant material in the greenhouse or in the field, the original position of the explant in the plant, the time of the year, the endogenous level of hormones, the size of the explant, the method of inoculation, the nutrients available, the regulators, the physical growth factors, addition of other substances to the medium etc.

Considering the complexity of regeneration as shown above it is obviously impossible to give a review of all the relevant literature. The writing of such a review is also complicated by the following facts:

1. The literature is so extensive that it is practically impossible to condense; most publications only describe the conditions under which regeneration takes place, and how these may be achieved. The question as to why regeneration takes place has not so far been answered (Street, 1979).
2. It is not possible to generalize, since each plant species, and in some cases individual cultivars, have their own requirements.
3. Some plants have such a high or such a low regenerative capacity that organ formation is difficult to regulate. Endogenous factors then predetermine the reaction (whether positive or negative). Exogenous factors (such as regulators) normally have little effect in these cases.
4. Interaction of different factors (e.g. light and cytokinin; darkness and temperature; auxin and cytokinin, etc.) make the overall picture very complicated. A good example of the interaction between auxin, light and pH, is shown by the work of Williams et al. (1985) using Australian woody species.
5. If one or more of the factors are limiting for adventitious organ formation, then few conclusions can be made about the other factors.
6. Root formation usually opposes shoot formation and vice versa (although this is not always the case!). If both processes are attempted simultaneously in vitro, then problems often arise. To obtain whole plants it is best to initially induce adventitious shoot formation and subsequently induce adventitious root formation.
7. There is an almost total lack of fundamental knowledge concerning the important role played by regulators during regeneration. In fact little is known about the working mechanism of regulators at a molecular level (Street, 1979).
8. Little is understood concerning the difference between the juvenile and adult phase, which are of such importance in regeneration.

Sometimes as a result of the adult phase of a plant it is almost impossible to induce adventitious organ formation in vitro; a good example of this being the explant culture of adult gymnosperms. David (1982) and Bonga (1982) gave some advice on how to make regeneration possible in such cases:

1. Isolate the explant from the basal part (juvenile zone) of a tree. Regeneration is then sometimes possible (Fig. 20.9), an example being *Sequoia sempervirens.*
2. Graft a meristem or shoot tip of an adult plant onto a seedling to obtain rejuvenation.
3. Use a rejuvenated shoot, which can arise on roots, on a wound of a pollarded tree; or use sphaeroblasts (Fig. 20.9).

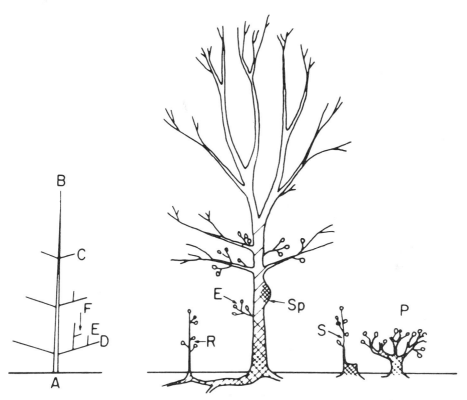

Fig. 20.9. Diagram of juvenility gradients in trees. Left: juvenility in a regularly shaped conifer. The degree of juvenility of an apical meristem is inversely proportional to the distance (along trunk and branches) between the root-shoot junction (A) and the meristem. Since laterals are shorter than the internodes from which they originate when considering the distances AB AC AD AE AF, meristem B is therefore the most mature, and meristem F is the most juvenile. Right: juvenility in several of the hardwoods. Density of crosshatching indicates degree of juvenility. Epicormic shoots (E), spheroblasts (Sp), root shoots (R), stump shoots (S), and severely pruned trees (P) are juvenile. In the juvenile zone, note the single trunk, retention of leaves close to the trunk in winter, and obtuse branch angles. In the mature zone, note the forked trunk and acute branch angles (Bonga, 1982).

20.4.2. *Adventitious root formation*

In this Chapter a review is given of the factors which determine adventitious root formation in the single-node method and with axillary shoots and explants. No detailed information is given in this Chapter concerning the establishment of in vitro formed shoots in vivo, since this has been adequately dealt with in Chapter 14; the fact that this method of rooting is becoming important can be seen from the work of Maene and Debergh (1983), who transplanted in vitro produced shoots successfully in vivo, for *Philodendron, Cordyline terminalis, Begonia* and *Ficus benjamina.* This method has also become important in many countries for gerbera shoots.

The overview below is mainly compiled from the work of Gautheret (1969), Pierik and Steegmans (1975a), Pierik et al. (unpublished data) and Németh (1986).

Age and developmental stage of the plant
In most cases root formation is far more easily induced in juvenile than in adult plants, although the reasons for this are usually not clear (Hackett, 1970). Vegetative plants are more responsive than flowering plants, although there are several exceptions to this general rule. Shoots regenerate more easily when taken from the basal part of a tree, this being linked to the juvenile character of the base of a tree. The regenerative capacity of cotyledons is usually much higher than that for the leaves formed later. The first pair of leaves of a (rejuvenated) shoot regenerate more rapidly than subsequent leaves.

However, although vegetative plants usually regenerate better than generative ones, many exceptions have been found: *Freesia* (Bajaj and Pierik, 1974), *Lunaria annua* (Pierik, 1967), tobacco (Paulet, 1965), *Nerine bowdenii* and *Eucharis grandiflora* (Pierik and Steegmans, 1986a), *Musa* (Cronauer and Krikorian, 1986) and *Phoenix dactylifera* (Drira et al., 1986).

Position on the plant from which the explant is taken
Examples are known where in principle each segment from a particular plant organ has approximately the same regenerative capacity. If differences are present, they can usually be explained by the fact that for example within a leaf (stalk) there are tissues of different ages. On the other hand, large differences in the regeneration capacity within an organ have been established: e.g. in the flower stalk of the biennial honesty there are two regeneration gradients (a root-forming and a shoot-forming gradient) which are opposite to each other (Pierik, 1967). This sort of regenerative gradient has also been shown for tobacco stalks (Aghion-Prat, 1965) and bulb scales of lily (van Aartrijk, 1984) and hyacinth (Pierik and Ruibing, 1973). The basal explants from these bulb scales regenerate far better than the terminal explants. Positional (topophysical) effects on regeneration are also common in trees and shrubs, and cyclophysical (same age but different position) effects and periphysical (same age and position, but different exposure to e.g. sun or shade light) effects may also play a role.

Plant species and cultivars

As was mentioned in the production to this Section regenerative capacity differs between plant species. Herbaceous plants regenerate far easier than trees and shrubs. The fact that different cultivars of herbaceous plants sometimes have different regenerative capacities is just as hard to explain as the differences between species. Sometimes regenerative differences are explained by the presence or absence of an anatomical barrier, as is the case for *Rhododendron;* wounding can often have a positive effect in such cases.

Sex

It was shown in a number of plant species *(Actinidia chinensis, Populus ciliata)* that female plants have a greater regenerative capacity than male plants (George and Sherrington, 1984).

The size of the explant

Larger explants are sometimes easier to regenerate than smaller ones, perhaps due to the presence of more food reserves. On the other hand, small explants have relatively larger wound surface, which have been shown to promote regeneration in the case of lily scales (van Aartrijk, 1984).

Wounding

It has been established by Pierik and Steegmans (1975a) for *Rhododendron* and by Snir and Erez (1980) for apple rootstocks that rooting is promoted by wounding.

Orientation of inoculation

This plays an important role with many plants. Apolar inoculation (upside down), promotes and polar inoculation (natural orientation, base down) inhibits regeneration. This is perhaps due to a better O_2 supply above the medium, but can also be linked to accumulation phenomena (Pierik and Steegmans, 1975a).

Number of subcultures

Adventitious root formation on repeatedly subcultured shoots of 6 members of the *Rosaceae* declined with subculturing on all species tested, with the exception of *Chaenomeles*. This decline in rooting potential was less rapid on media containing 2iP than on BA (Norton and Norton, 1986a). However, in shoot cultures of the difficult-to-root apple cultivar Jonathan, Noiton et al. (1986) showed that with increasing subculture, rooting percentage increased from 0 in the initial explants to 100 after 9 subcultures. This can perhaps be explained on the basis of the rejuvenation that takes place, especially with woody plants with successive subcultures.

Oxygen supply

As has already been mentioned the O_2 supply plays an important role. This is one reason why shoots often root better in vivo than in vitro, where there is a limited supply of O_2 in the agar (O_2 is not very soluble in water). Roots formed in an agar medium often have poorly formed root-hairs, due to lack of O_2. Aeration often promotes the formation of roots, and rooting is also better with a liquid medium.

To promote better elongation of shoots and to bring about improved rooting (without transferring onto a rooting medium) Maene and De-bergh (1985) added liquid media to exhausted solid media.

Light

Light generally has a negative effect on root formation. Plants which have been grown in the dark (etiolated), root more easily than light-grown plants such as e.g. *Prunus cerasifera* (Hammerschlag, 1982). A few plants such as *Helianthus tuberosus* and lily react positively to light. Higher irradiances inhibit root formation more than lower irradiances, although gerbera is an exception to this rule (Murashige et al., 1974). It is as yet not known why light has an inhibitory effect on root formation (Pierik and Steegmans, 1975a).

Economou and Read (1986) showed with *Petunia* hybrids that shoots grown under red light produced more and heavier roots than those from far-red treated cultures; this effect of red light on micro-shoot rooting is reversible by a subsequent exposure to far-red light.

Temperature

Adventitious root formation is generally promoted by high temperatures. Only *Lunaria annua* (Pierik, 1967) and some woody plants (Anonymous, 1978a) have a cold requirement for root formation. The time of isolation can also have a great influence, especially with woody plants: it is more difficult to promote growth and organ formation in dormant plants (having received no cold treatment to break dormancy) than in non-dormant plants.

Agar

Little is known about the effect on root formation of agar and it's concentration. A short literature review is given by Németh (1986). The formation of adventitious roots, especially of woody plants, is generally poor on solid media; for this reason a liquid medium is sometimes used

206

for woody plants. Moreover, the roots can also grow upwards on a solid medium.

Sugar

It is firmly established that sugar is necessary for adventitious root formation. Since the most efficient root formation takes place in the dark, it is understandable why sugar plays such an important role. However, this does not explain everything since light and sugar are not mutually replaceable (Evers, 1984). It is striking that a high sugar concentration is needed for adventitious root formation, especially in the case of woody plants; an exception to this is *Pseudotsuga menziesii* which requires a low sugar concentration (Cheng and Voqui, 1977).

Osmotic value

Hardly anything is known about the effect of the osmotic value on root formation. Pierik and Steegmans (1975a) showed with *Rhododendron* that root formation was prevented by increasing the osmotic value from 2.5 to 6.0 bar.

pH

Little is known about the effect of pH on root formation, although a few *Prunus* species prefer a low pH (Németh, 1986).

Mineral nutrients

The effect of mineral nutrients on root formation is also little understood. The need for macro- and micro-elements is strongly dependent on the amount of food reserves available in the explant. The fact that the mineral nutrients can have an important role has been shown by Tripathi (1974). Woody plants, and in particular gymnosperms prefer a low salt concentration for the formation of adventitious roots (Anonymous, 1978a). More information about this subject has been given by Németh (1986).

Auxin

Most plants require the presence of auxins for efficient root regeneration. This need is not constant since after root initiation (high auxin requirement), outgrowth of the root primordia requires a low concentration (Fig. 20.10); for this reason sometimes root initiation is carried out in vitro under a high auxin concentration, and the outgrowth of root primordia takes place in vivo (Staritsky, personal communication). The weak auxin IAA is often used during rooting of herbaceous plants; in time IAA is broken down in light which means that the root primordia produced can then develop better. Woody plants require higher auxin concentrations than herbaceous plants. The most efficient auxins are definitely IBA and NAA. In some cases a mixture of auxins is added (IBA and NAA). IAA is normally not very efficient when used for plants which are difficult to root (Pierik and Sprenkels, 1984). The formation of excessive callus due to the influence of very strong auxins, such as 2,4–D and NAA, should be avoided.

Other regulators
Adventitious root formation is generally inhibited by cytokinins, gibberellins and abscisic acid. In some cases cytokinin may have a stimulative effect: *Prunus* and *Cydonia* (Németh, 1986) and *Dalbergia sissoo* (Datta et al., 1983). In a few cases GA_3 also has a promotive effect. To induce adventitious root formation after axillary shoot propagation cytokinin is usually left out and auxin added. Little is known of the effect of ethylene on root formation. In a controversial piece of research Coleman et al. (1980) proposed that in the case of tomato explants ethylene had an

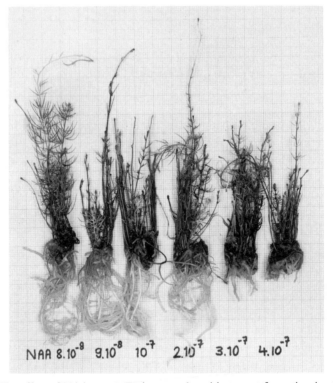

$$NAA\ 8.10^{-8}\quad 9.10^{-8}\quad 10^{-7}\quad 2.10^{-7}\quad 3.10^{-7}\quad 4.10^{-7}$$

Fig. 20.10. The effect of NAA concentration on adventitious root formation in *Asparagus officinalis*. Elongation of the adventitiously formed roots decreases with increasing NAA concentration. Photograph taken 16 weeks after isolation.

inhibitory effect on root formation in the presence of IAA; if the ethylene produced by the plant was removed then root formation took place under the influence of the IAA. Considering the interesting research carried out by van Aartrijk (1984) with lily scales and the central role played by ethylene in bulb regeneration, this topic will certainly receive more attention in the future.

Recently the oligosaccharides have been discovered as a new group of substances with regulatory functions (Albersheim and Darvill, 1985). These substances appear to influence organ formation in tobacco in much the same way as cytokinins and auxins.

Diverse substances
Cheema and Sharma (1983) and Preil and Engelhardt (1977) showed that active charcoal can play an important role in the formation of adventitious roots. It can have two effects: the adsorption of all organic compounds with the exception of sugars, no light supply to the nutrient media.

Phloroglucinol added together with IAA sometimes has a synergistic effect on adventitious root formation and may be added in the case of woody plants (Anonymous, 1978a). From the research carried out by Jones and Hopgood (1979), it appears that phloroglucinol stimulates root formation of in vitro grown shoots of *Malus*. However, the claims of these authors have been challenged in the literature.

20.4.3. *Adventitious shoot formation*

Adventitious shoot formation is used, in practical horticulture for a number of plants as a means of vegetative propagation; e.g. *Saintpaulia, Begonia, Achimenes, Streptocarpus*, lily, hyacinth and *Nerine*, are now cloned in large numbers using this method.

The fact that adventitious shoot formation is by no means as popular as the axillary shoot method is due to the following:

1. The number of plants that can form adventitious shoots in vivo and/or in vitro is much smaller than those which are capable of forming adventitious roots.
2. The chances of mutants arising with this method are far higher than with the single-node or the axillary shoot methods.

There are considerable similarities between adventitious shoot and root formation: factors which have a positive influence on root formation often positively promote shoot formation. This is illustrated below with a few examples:

1. Juvenile plant parts regenerate adventitious shoots relatively easily. For example with gymnosperms, adventitious shoot formation is often only possible with embryos, seedlings or parts of seedlings and not with plants (David, 1982; Anonymous, 1984). In the case of trees true ontogenetic rejuvenation is basic for multiplication.
2. Sugar always plays a very important role in organ (root and shoot) formation.
3. Root and shoot formation are usually both inhibited by gibberellins and abscisic acid.

On account of the points discussed above only those factors which have a different effect on adventitious shoot and root formation, will be considered in this Section.

Adventitious shoot formation is promoted by light in many cases: the role played by the light quality and irridiance is discussed in Section 13.2. It is noteworthy that there are also a number of plants which require darkness for adventitious shoot formation: flower buds of *Freesia* (Pierik and Steegmans, 1975b), peduncle explants of *Eucharis grandiflora* and *Nerine bowdenii* (Pierik and Steegmans, 1986a; Pierik et al., 1985). Economou and Read (1986) showed that leaf fragments of *Petunia hybrida* grown on a cytokinin-free medium and treated with red light, produced more shoots which had a greater fresh weight than explants from far-red treated cultures. However, on a medium with BA, red and far-red light treatments gave similar number of shoots and fresh weight, which were greater than those in cultures that were cytokinin-free.

Although a high temperature is often necessary for adventitious shoot formation, there are a few exceptions to this rule: *Begonia* (Heide, 1965) and *Streptocarpus* (Appelgren and Heide, 1972).

Unlike root formation which is often promoted by apolar inoculation this is not always the case with adventitious shoot formation. Pierik and Zwart (unpublished) demonstrated with leaf explants of *Kalanchoë farinacea* that shoot regeneration is strongly promoted when the lower epidermis of the flat-inoculated explant lies on the medium; it has been conclusively shown that flat rather than vertical inoculation produces better shoot regeneration.

The auxin and cytokinin requirements for adventitious shoot formation are somewhat complicated:

1. There are plants which in principle need neither cytokinins nor auxins for the formation of adventitious shoots. Examples are the biennial honesty (Pierik, 1967), chicory (Pierik, 1966), *Cardamine pratensis* (Pierik, 1967a), *Streptocarpus* (Appelgren and Heide, 1972; Rossini

and Nitsch, 1966). However, this is not to say that the addition of cytokinins and/or auxins does not have a positive or negative effect on shoot formation.

2. Most plants require cytokinin for shoot formation, whereas auxin generally prevents it (Miller and Skoog, 1953; Paulet, 1965; Nitsch, 1968).

3. There is a further group of plants which require exogenous auxin for shoot formation: lily (van Aartrijk, 1984), hyacinth (Pierik and Steegmans, 1975d).

4. It appears that a high cytokinin concentration and a low auxin concentration are very important for shoot formation in many different plant species (Fig. 20.11). A few examples are: *Begonia* (Ringe and Nitsch, 1968; Heide, 1965), horseradish (Wurm, 1960; Sastri, 1963), foxglove (Dolfus and Nicolas-Prat, 1969), *Atropa belladonna* (Zenkteler, 1971) and cauliflower (Margara, 1969).

It can be concluded that in those plants which have a cytokinin and auxin requirement, the cytokinin concentration should usually be high and that of auxin low; the ratio between these two regulators determining organ formation (Skoog and Tsui, 1948; Miller and Skoog, 1953; Paulet, 1965; Gautheret, 1959). The cytokinin BA, is easily the most efficient in promoting adventitious shoot formation in many plants and is used

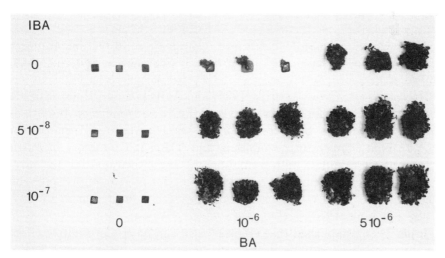

Fig. 20.11. The influence of various IBA and BA concentrations on adventitious shoot formation of leaf explants of *Kalanchoë* 'Tessa'. BA is required for adventitious shoot formation, whereas IBA is not. However, BA and IBA together have a synergistic effect on adventitious shoot formation (Pierik and van Marrewijk, unpublished results). Photograph taken 8 weeks after isolation.

exclusively in the case of gymnosperms. Other cytokinins (kinetin, 2iP, PBA and zeatin) may be used but, in practice, less frequently. It is often difficult to predict which cytokinin will be the most suitable. When using BA it should be borne in mind that, especially when using high concentrations, anomalies (miss-formed and abnormal shoots) can arise. It is often advisable to transfer adventitious shoots, which have been formed on a medium with a high cytokinin concentration, to one with a low cytokinin concentration to promote further outgrowth and development.

In some cases a combination of cytokinin and adenine (sulphate) is favourable for shoot formation (Skoog and Miller, 1957; Nitsch et al., 1969a; Nitsch, 1968; Dolfus and Nicolas-Prat, 1969). Adenine (sulphate) alone may be enough in some cases to induce shoot formation: *Begonia* (Ringe and Nitsch, 1968), tobacco (Skoog and Tsui, 1948; Paulet, 1965), elm (Jacquiot, 1951), *Plumbago indica* (Nitsch, 1968). No positive effect of adenine (sulphate) was found on adventitious shoot formation with *Kalanchoë farinacea* leaf explants by Pierik and Zwart (unpublished data).

Usually an increase in the gibberellin concentration stops adventitious shoot formation: hyacinth (Pierik and Steegmans, 1975d), biennial honesty (Pierik, 1967), chicory (Paulet, 1965; Bouriquet and Vasseur, 1965), tobacco (Aghion-Prat, 1965; Murashige, 1961, 1964), *Streptocarpus* (Ringe and Nitsch, 1968; Appelgren and Heide, 1972), *Begonia rex* (Schraudolf and Reinert, 1959; Reinert and Besemer, 1967; Bigot and Nitsch, 1968), *Plumbago indica* (Nitsch, 1968).

Abscisic acid normally inhibits adventitious shoot formation, although it induces it in the case of *Ipomoea batatas* (Yamaguchi and Nakajima, 1974).

Recent research (van Aartrijk, 1984; van Aartrijk et al., 1986) with lily scale explants, has shown that ethylene (formation) plays an important role in adventitious bulb formation. It was demonstrated that auxin (NAA), high temperature, wounding and TIBA (an inhibitor of auxin transport) all stimulate bulb formation. All these factors influence ethylene biosynthesis. A positive effect of antiauxins (N-1-naphthyl-phthalic acid) has also been shown on adventitious shoot formation in tobacco (Feng and Link, 1970).

It has been established that vitamins also can have a promotive effect on adventitious shoot formation: cyclamen (Mayer, 1956; Stichel, 1959), elm (Jacquiot, 1951), horseradish (Wurm, 1960), *Robinia* (Seeliger, 1959).

The ability to change, with the help of *A. tumefaciens,* the production of plant hormones and thereby make adventitious shoot formation possible will certainly be a of great importance in the future. In this way plants which normally are not able to form adventitious shoots for example as a result of the adult state (especially with treees and shrubs), are then able to do so. This can lead to rejuvenation (See Chapter 11), which is particularly important for vegetative propagation. A few interesting examples of adventitious shoot formation as a result of infection by *A. tumefaciens* were given by Ooms et al. (1983) for potato and other plants. Fillati et al. (1986) demonstrated that the wild type *A. tumefaciens* strain C58 enhanced the regeneration ability of *Populus* as a result of transformation. *A. tumefaciens* has also been used to obtain adventitious shoot formation in *Nicotiana tabacum* (Steffen et al., 1986).

20.5. Callus induction, callus culture and regeneration of organs and embryos

20.5.1. *Introduction*

Callus is basically a more or less non-organized tumour tissue which usually arises on wounds of differentiated tissues and organs. The intiation of callus formation is referred to as callus induction. It is possible with many different plant species to produce a callus and then to grow this further on a new medium; this further growth is termed callus subculture. In exceptional conditions, and sometimes spontaneously, the regeneration of adventitious organs and/or embryos can occur from a callus. In a liquid medium a callus may form aggregates (clumps of cells) or individual cells. Plants can even regenerate from individual cells.

A schematic illustration of the entire process from isolation of an explant to the regeneration of a plant from a callus and clumps of cells is given in Fig. 20.12. A young leaf is removed from a plant and after sterilization cut into explants. In tube 1 a leaf explant is isolated on a solid medium in order to obtain callus tissue on the wound surfaces. This is realized in the second tube. Callus subculture on a solid medium is carried out in tube 3; the callus formed in tube 2 is cut from the explant and inoculated in tube 3. The callus is strongly grown in tube 4. Further propagation after tube 4 can take place on a solid medium, but can also be carried out in a liquid medium (Erlenmeyer flasks placed on a shaker). In flask 5a a few callus clumps are inoculated which propagated themselves within a few weeks (flask 6a). If needed, propagation in a liquid medium can take place repeatedly. If regeneration of organs or embryos from the callus is required at a particular time, then the callus or the cell suspension should be inculated onto another medium. Organ formation

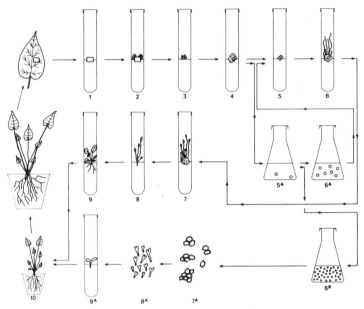

Fig. 20.12. Schematic illustration of callus induction, callus culture, organ and somatic embryo regeneration, and plant formation. For further explanation see text.

is normally carried out on a solid medium; callus tissue from tube 4 or flask 6a is inoculated onto an organ-forming medium in tube 5. In tube 6 the adventitious shoots formed from the callus can be seen. Although embryos can arise from calluses grown on a solid medium, embrogenesis is more often occurring in a liquid medium as shown in 7a and 8a. The embryos are grown into plants on a solid medium in tube 9a and then transferred into soil for further development (10). When adventitious shoots develop on the solid mendium in tubes 6 and 7, shoot cuttings are taken in tube 8, rooted in tube 9 and ultimately transferred to soil (10).

There is so much literature concerning callus induction, callus culture, cell culture, organogenesis and embryogenesis that it is impossible to give a concise picture in this book. In this Chapter the main points will be considered. The following literature has been used for compiling this Chapter: Street (1973, 1979), King and Street (1973), Thomas and Davey (1975), Thomas et al. (1979), Tisserat et al. (1979), Kohlenbach (1978), David (1982), Sharp et al. (1982) and Evans et al. (1981).

20.5.2. *Callus induction*

If there are only differentiated cells present in an isolated explant, then de-differentiation must take place before cell division can occur; parenchyma cells usually undergoing this de-differentiation (Halperin, 1969). If the explant already contains meristematic tissue when isolated, then this can divide immediately without de-differentiation taking place. De-differentiation plays a very important role, enabling mature cells in an explant isolated from a adult plant to be redetermined. In this process adult cells are (temporarily) able to revert from the adult to the juvenile state (rejuvenation). This rejuvenation can have very important consequences: cells are induced to divide intensively; rejuvenated cells have a greater growth and division potential than adult cells; rejuvenated cells are able, under special circumstances, to regenerate into organs and/or embryos.

After de-differentiation the cells may begin to divide intensively under the influence of such factors as regulators present in nutrient medium; a tumour tissue is then formed, which is known as a callus. Callus is, in principle, a non-organized and little differentiated tissue; however differentiated tissues can be present especially in larger clumps of callus tissue. When critically examined, callus culture can be seen to be far from homogeneous, because it is usually made up of two types of tissues: differentiated and non-differentiated.

All types of organs (roots, stems, leaves, flowers, etc.) and tissues can be used for starting material for callus induction. If the callus is difficult to induce, or if juvenile callus is needed then (immature) embryos or seedlings or parts of these are used. It should be taken into account that the type of starting material (juvenile or adult, and the original position of the explant in the plant (reflects the endogenous hormone level) can both have an important influence on such processes as cell division and organ and embryo formation.

Monocotyledons react differently when considering callus induction generally being less likely to form callus tissue than dicotyledons; it is often only necessary to add auxin as the hormonal stimulus for callus induction (Fig. 20.13). Since it is sometimes difficult to form callus tissue with monocotyledons embryos, young leaves, seedlings or very young flower initials are often resorted to as starting material.

An exogenous supply of regulators is often recommended to initiate callus formation on an explant. The exogenous regulator requirement (type of regulator, concentration, auxin/cytokinin ratio) depends strongly on the genotype and endogenous hormone content. These requirements can, in principle be split into three types:

1. Auxin only needed (especially monocotyledons).
2. Cytokinin alone required.
3. Both auxin and cytokinin required.

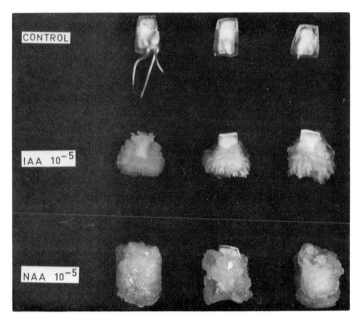

Fig. 20.13. Only adventitious roots and bulblets are formed from bulb scale explants of hyacinth in the absence of auxin (control). An addition of 10^{-5} g ml^{-1} IAA induces callus as well as root formation, whereas the addition of 10^{-5} g ml^{-1} NAA only induces pronounced callus formation. Photograph taken 8 weeks after isolation.

Before the discovery of the cytokinins it was already appreciated that a combination of the auxin 2,4-D and coconut milk was extremely effective in inducing callus formation; this was demonstrated with carrot by Steward et al. (1964). After the discovery of the cytokinin kinetin is was shown that in most, although not in all cases, coconut milk could be replaced by cytokinin. The exceptions are due to the fact that coconut milk has many other components apart from cytokinin. The fact that auxin and cytokinin are essential for callus induction was fully appreciated after the discovery of the presence of cytokinin in coconut milk.

Many other factors, as well as regulators, are important for callus formation: genotype, composition of the nutrient medium, physical growth factors (light, temperature, etc.). The Murashige and Skoog (1962) mineral medium, or modifications of this are often used. Saccharose or glucose 2–4% are usually employed as the sugar source. If the exact nutrient requirements for callus formation are unkown, then other compounds may be added to the medium: casein hydrolysate, malt extract, yeast extract or coconut milk. The effect of irradiance on callus formation is dependent on the plant species; light may be required or promotive in

216

some cases and darkness in other cases. A high temperature (22–28 °C) is normally advantageous for callus formation.

20.5.3. *Callus culture*

20.5.3.1. *Introduction.* After induction the callus is grown further on a new medium. The first subculture is usually carried out onto a solid medium, although in some cases it is possible to begin directly with a liquid medium. If the callus has poor growth during the first subculture then it may be necessary to also subculture a part of the explant. The ngrowth conditions (nutrient medium, physical growth factors) are usually similar to those for callus induction, only the auxin and cytokinin concentrations are normally lower. If the growth of the callus stagnates after subculturing then this indicates that the subculture medium is not suitable. Sometimes further growth of the callus is not possible on synthetic media and then complicated mixtures of substances may be added: coconut milk, casein hydrolysate, malt extract, yeast extract, etc.

Callus tissue from different plant species may be different in structure and growth habit: white or coloured, free (easily separated) or fixed, soft (watery) or hard, easy or difficult to separate into cells and aggregates in liquid media. The callus growth within a plant species may also vary depending on such factors as the original position of the explant within the plant, and the growth conditions. On the other hand it is sometimes possible to see differences in colour and structure within a callus that are not due to these factors and then the possibility of mutation should be considered.

Habituation may play a role in callus culture: a callus may loose it's auxin and/or cytokinin requirements after repeated subculture; the callus is then autotrophic with regards to regulators. How much this is due to mutation or epigenetic change is strongly contended in the literature.

Callus is understood as a rapidly growing tissue which is relatively easy to grow. If plants are to be propagated vegetatively by the means of callus culture then the callus should fill the following requirements:

1. The regenerative potential should not be lost after repeated subculture. It appears that the potential of a callus to form organs and embryos can be lost (Thomas and Street, 1970), as was shown in the case of carrot (Syono, 1965a, 1965b, 1965c).
2. Genetic stability: the material obtained should be identical to the starting material. This is definitely not always possible and more details of this problem are found in Chapter 21.

20.5.3.2. *On solid media.* In older literature callus culture is almost entirely described on a solid medium, since liquid media had not been developed. Callus culture is generally much slower on solid rather than liquid media which can be shaken. The reasons for this difference in growth rate are:

1. The callus clump has only limited contact with the solid medium and therefore takes up less compounds from the medium.
2. Oxygen supply is better in liquid media.
3. Callus clumps are able to fragment in liquid media producing a greater surface area with a resulting improvement in uptake of material from the medium.
4. Gradients are present on a solid medium which can cause earlier differentiation, inhibiting the callus growth.

20.5.3.3. *In liquid media.* Erlenmeyer flasks placed on a rotary shaker have become extremely popular for callus propagation. The growth and growth conditions for callus culture in liquid media are described in detail in the handbook by Street (1973). Rather than attempting to summarize the vast literature concerning callus culture in liquid media a few general rules are given below:

1. Rotary shakers are generally used to prevent the liquid growth media from being lost from the flasks at high speeds.
2. The flask should not be more than 20–30% full.
3. The amount of medium in the flask can normally be reduced if less cells or clumps of cells are inoculated.
4. The optimal revolution of the machine should be about 80–100 r.p.m., although in some cases lower (40) and higher (120) speeds may be used.
5. The growth of cells, callus etc. generally takes place as follows: initially (the lag phase) there is practically no growth. This increases exponentially, then becomes linear before slowing down and finally stopping (Street, 1973).
6. The rapid establishment of cells in suspension cultures is proportional to the cell density at the beginning of the culture.

More information on growth in liquid media (batch cultures, renewal of media, continuous culture, interim removal of cells) can be found in Chapter 11. Whether the culture in liquid media is referred to as a suspension culture (a mixture of separate cells and cell aggregates) or callus clump culture strongly depends on the plant species and the type of cell under consideration: *Anthurium andreanum* callus grows in clumps in liquid media (Pierik and Steegmans, 1975c), and is very difficult to obtain in suspension (Fig. 20.14). In contrast carrot callus very easily separates into cells and aggregates of cells. Suspension cultures, as well as callus cultures, are morphologically and/or cytogenetically heterogeneous.

20.5.4. *Regeneration of organs and embryos*

20.5.4.1. *Introduction.* Depending on the plant species and where the callus originated, adventitious organ and/or the formation of so-called somatic embryos can take place. Callus tissue originating from juvenile and/or herbaceous material generally regenerates much better than material from adult and/or woody plants. In the case of gymnosperms, virtually only juvenile material can be used for regeneration (David, 1982). Thomas et al. (1979) give a long list of plants in their review article which can regenerate plants from protoplasts, cells or callus tissue. A larger number of woody plants are also able to regenerate e.g. *Coffea canephora* (Staritsky and van Hasselt, 1980; Pierson et al., 1983), *Ilex aquifolium,*

Fig. 20.14. Callus of *Anthurium andreanum* grown in liquid medium continues to grow in clumps. Left: a genotype where the callus fragments; right: a genotype where the fragmentation is slight. Photograph taken 12 weeks after isolation.

Coffea arabica, Corylus avellana, Populus species (Pierik, 1979), *Citrus species* (Litz, 1985).

As was mentioned earlier the regenerative capacity of cells and callus tissue may reduce or be completely lost if growth is continued for too long. The reason for this is so far unknown, as to the reason why the calluses of some plants are only capable of forming adventitious organs and no embryos whereas others behave entirely opposite. Some plant species are capable of regenerating by either adventitious shoots or adventitious embryos (Thomas and Davey, 1975).

It is not easy to give fast guidelines for organ regeneration. In some cases regulators play the main role in regeneration, but in others regeneration may start spontaneously or may be induced by the absence of regulators (Reinert, 1973).

20.5.4.2. Regeneration of organs. A callus is usually more capable of regenerating adventitious roots than adventitious shoots (Thomas and Davey, 1975). Root formation generally takes place in a medium with a relatively high auxin and low cytokinin concentration (Fig. 20.15); this is in agreement with regeneration of explants described in Section 20.4. The initiation of root primordia often requires a higher auxin concentration

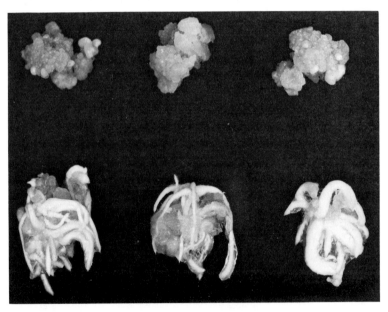

Fig. 20.15. Above: callus culture of *Freesia* 'Ballerina' on a nutrient medium with 5 mg l^{-1} PBA and 0.1 mg l^{-1} NAA (no organ formation). Below: culture on a medium with 0.5 mg l^{-1} PBA and 1 mg l^{-1} NAA (abundant root formation). From Bajaj and Pierik (1974). Photograph taken 12 weeks after isolation.

220

than that required for the outgrowth of root primordia. Roots and shoots are usually formed completely independently from each other i.e. there is no direct link between them if they arise from the callus at the same time.

Adventitious shoot formation can appear in callus tissue if there is a low auxin concentration and a high cytokinin concentration. BA is the most effective cytokinin for inducing the formation of adventitious shoots. In some cases adventitious roots are formed at the base of the adventitious shoots.

The induction of adventitious shoots on callus tissue is perhaps influenced by far more complex factors than so far appreciated. In some cases a high salt concentration (which inhibits growth?) can induce regeneration. It is certain that raising the ammonium concentration can promote organogenesis; Pierik and Steegmans (1976) showed that with *Anthurium andreanum* a lowering of the ammonium concentration promoted the formation of adventitious shoots on callus tissue (Fig. 20.16).

Quite recently, Rathore and Goldsworthy (1985) showed that electric currents of the order of 1–2 µA stimulate shoot differentiation up to 5-fold in *Nicotiana tabacum* callus cultures; they also observed that root and shoot growth were promoted by electric currents. It has been sug-

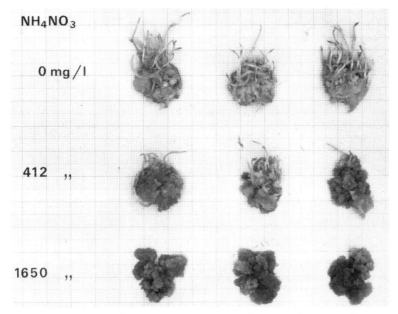

Fig. 20.16. The promotion of adventitious shoot formation in *Anthurium andraeanum* callus, by low levels of ammonium nitrate in the culture medium (Pierik, 1976a). Photograph taken 14 weeks after isolation.

gested that the effect may be due to the artificial currents controlling the natural currents, which flow through plant cells.

20.5.4.3. *Regeneration of embryos.*

When embryos regenerate from somatic cells or tissues (which are haploid, diploid etc.), it is termed somatic embryogenesis; which is the opposite to zygotic embryogenesis which results from the fertilization of an egg cell. In the literature, somatic embryos are referred to by many names: embryo-like structures, adventitious or vegetative embryos, embryoids; and the process is called adventitious, asexual or somatic embryogenesis (Sharp et al., 1982).

In 1958 Reinert and Steward showed somatic embryogenesis for the first time with callus and suspension cultures of *Daucus carota* (carrot). Steward (1958) discovered that the in vitro development of the somatic embryo had great similarities with zygotic embryogenesis which takes place in the liquid endosperm; he also proposed that the expression of totipotency appeared by isolation (spatial as well as physiological) of a cell or group of cells. He saw a parallel between the isolation of a fertilized egg cell, and that of a somatic cell in vitro. The promotive effect on somatic embryogenesis of the liquid endosperm of the coconut (coconut milk) fits well with this concept: the fertilized egg cell is situated in the middle of the endosperm which is it's nutrient (Steward et al., 1964). Together with the carrot other members of the *Umbelliferae* (celery and caraway) were classically used to study somatic embryogenesis in vitro (Thomas et al., 1979). Although Steward (1958) used coconut milk for the induction of somatic embryogenesis in carrot, Reinert (1959) showed that it could also be satisfactorily induced on a synthetic medium. Despite the many similarities between zygotic and somatic embryogenesis there are also obvious differences (Tisserat et al., 1979).

The development of somatic embryos and plantlets from an explant (Fig. 20.17) can be summarized as follows (Kohlenbach, 1977). Differentiated cells should first be de-differentiated, after which they start to divide. In this way a non-organized mass of vacuolated parenchyma cells arises. This is then transformed into cytoplasma-rich cells which become embryogenic under the influence of auxin. The embryogenic cells from which embryoids are visually derived show a number of common features, which are characteristic of rapidly dividing cells. These include, small size, dense cytoplasmic contents, large nuclei with prominent enlarged nucleoli, small vacuoles and a profusion of starch grains. Their histochemistry and ultrastructure are suggestive of intense RNA synthesis and metabolic activity (Williams and Maheswaran, 1986). The development of an embryo from the embryogenic cells is usually accomplished by the absence of auxin in the medium.

222

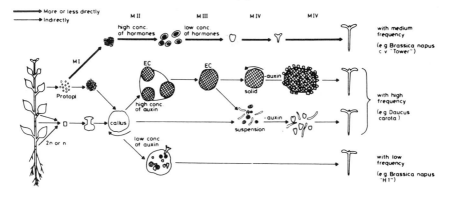

EC Embryogenic Clump
● With Embryogenic Determination
○ Somatic Embryo

Fig. 20.17. Schematic illustration of somatic embryogenesis (Kohlenbach, 1985). In the upper row (MI-MIV) somatic embryos are formed more or less directly (without an intervening callus phase) from embryogenically non-determined leaf protoplasts (protopl.); the embryos are induced by a high concentration of hormones (MII), whereas embryo outgrowth (MIII) is obtained by a low concentration of hormones; MI-MIV are 4 steps, each with specific media.

In the lower part of the figure, callus (derived from protoplasts or from explant culture) is the starting point for embryogenesis; the auxin concentration regulates the induction of so-called embryogenic clumps (EC); embryogenic determination can be induced by high or low auxin concentration, or as a result of suspension culture; development of embryos from embryogenic clumps generally takes place by reducing or ommitting the auxin concentration.

The production of somatic embryos in cell suspensions and callus tissue (Kohlenbach, 1978; Tisserat et al., 1979) can take place either endogenously (inside) or exogenously (on the periphery); at the same time embryogenesis of nucellus tissue, hypocotyls of germinating plants, and of somatic embryos has also been described. Tisserat et al. (1979) and Sharp et al. (1982) found the following to be good starting material for somatic embryogenesis: parts of flowers, zygotic embryos, anthers, pollen grains and endosperm tissue. On the basis of the fact that embryos were seen to originate from nodular or bud-like masses of tissue, it was concluded that somatic embryogenesis was not unicellular in origin. After critical consideration it appears that somatic embryogenesis in fact takes place from a single cell, which segments and so forms a pro-embryo-like cell complex (Haccius, 1978). Zygotic, as well as somatic embryos, are then included in the definition given by Haccius (1978): a plant embryo is a new individual which arises from a single cell and has no vascular links with the mother tissue. This statement by Haccius that somatic

223

embryos are unicellular in origin has again come under dispute. For more discussion on this topic see Williams and Maheswaran (1986).

According to Ammirato (1983) it was already possible in 1979 to induce somatic embryogenesis in 132 plant species consisting of 81 genera and 32 families. Somatic embryogenesis is fairly common in the following families: *Ranunculaceae, Rutaceae, Solanaceae, Umbelliferae* and *Gramineae* (Tisserat et al., 1979).

Two different types of somatic embryogenesis (Fig. 20.17) can be differentiated (Sharp et al., 1982; Ammirato, 1983; Evans et al., 1981; Kohlenbach, 1985; Ammirato, 1986):

1. Direct: In this case an embryo arises directly from a cell or tissue without previous callus formation. The cells from which the embryo develops are called pre-embryonic determined cells (PEDC's). The starting material for this type of embryogenesis is already completely rejuvenated. Examples are nucellus tissues of *Citrus* species, which have a tendency for polyembryony (see Section 20.5.4.4.), and epidermal cells of hypocotyls (wild carrots, *Ranunculus sceleratus, Linum usitatissimum, Brassica napus 'Tower'*).

2. Indirect: In this type of embryogenesis a callus is formed from which embryos can form later. The cells from which the embryo arises are called embryogenically determined cells and form embryos when they are induced to do so (induced embryogenically determined cells; IEDS's). With this type of embryogenesis, differentiated cells must firstly be de-differentiated and then redetermined as embryogenic cells after cell division. Auxins (and sometimes cytokinins) are of great importance in this process. With this method, complete rejuvenation must first take place. Examples of this are: secondary phloem of carrot, leaf explants of *Coffea arabica, Petunia hybrida,* and *Asparagus officinalis.*

A few general rules for the induction of somatic embryos are given below (Thomas et al., 1979; Ammirato, 1983; Kohlenbach, 1978; Sharp et al., 1982; Tisserat et al., 1979; Raghavan, 1986).

1. A high auxin concentration is often required for embryo induction; for further development of the embryos this should be lowered or in some cases it should be completely eliminated from the medium. The auxin 2,4-D is very important for embryogenesis, especially in the case of grasses, examples of which are the cereal crops. Cytokinin does not have a vital role in this process.
2. Gibberellins and ethylene usually inhibit embryogenesis.

224

3. The chances of embryogenesis are greater with a callus that has originated from a juvenile plant. In the case of *Citrus* it appears that only nucellar tissue is very responsive.
4. Reduced nitrogen in the form of ammonium ions can be an important factor in embryogenesis. Potassium promotes embryogenesis, particularly if nitrogen is restricted. High calcium concentrations inhibit and high salt concentrations promote the formation of embryos.
5. Light generally promotes embryogenesis, although it can take place at low irradiance or even in darkness with some species.
6. High temperatures are normally favourable for somatic embryogenesis; some plants (especially anther cultures) require a cold shock to initiate the formation and further development (germination) of the embryos.
7. A saccharose concentration of about 2–3% is optimal.
8. Coconut milk often promotes embryogenesis (often quoted in early literature).
9. Abscisic acid, a naturally occurring plant growth inhibitor, exerts a number of striking effects on the somatic embryos grown in suspension culture (Ammirato, 1986a).

Somatic embryogenesis as a means of propagation is seldom used in practice, because:

1. There is a high probability of mutations arising.
2. The method is usually rather difficult.
3. With repeated subculture the chance of loosing regenerative capacity becomes greater.
4. Induction of embryogenesis is often very difficult or impossible with many plants species.
5. A deep dormancy often occurs with somatic embryogenesis which may be extremely difficult to break (Vilaplana-Marshall and Mullins, 1986).

Somatic embryogenesis has been carried out with a high degree of success with a number of plants. For example, the oil palm (Jones, 1983; Blake, 1983; Litz et al., 1985) is cloned at the moment on a large scale by callus, embryo and shoot formation. The oil palms so obtained appear to display no noteworthy genetic variability. Although polyploid cells are definitely found on occasion, it is certain that embryogenesis takes place exclusively from diploid cells. By no means all oil palm genotypes clone in vitro. Propagation in vitro has great advantages in the case of the oil palm which is impossible to clone in vivo. In this way plantations can be started with plants that have been selected for high productivity and resistance to diseases (Carantino, 1983); on the other hand multiclonal plantations of oil palm should also be borne in mind (Staritsky, personal communication). Considering the amount of research taking place at the moment it can be expected that methods will soon be found for in vitro propagation of other palms, such as the date palm, coconut palm, sago palm, which may be similar to that for the oil palm.

Somatic embryogenesis of haploid cells and tissues is dealth with fully in Chapter 23. Recent research on somatic embryogenesis is summarized in the book by Henke et al. (1985), and in the review article by Williams and Maheswaran (1986). Cell and protoplast culture and somatic embryogenesis of cereals is summarized in the book by Bright and Jones (1985).

20.5.4.4. *Nucellar polyembryony.* Polyembryony often occurs in vivo with many *Citrus* species: in a seed there is a zygotic embryo (arising from fusion of an egg cell and a male gamete) and also a number of embryos formed vegetatively (so-called adventitious embryos). These adventitious embryos regenerate from cells in the nucellus tissue and/or in the integuments, and are therefore referred to as nucellar embryos (Button and Kochba, 1977; Pierik, 1975; Rangan et al., 1968; Rangaswamy, 1981; Sharp et al., 1982). The nucellus tissue of *Citrus* can be described as a mass of juvenile tissue, which has a high regenerative capacity. Nucellar polyembryony has been described in 16 families to date (Rangaswamy, 1981).

In 1958 Rangaswamy (1981) discovered that in vitro isolated nucellus tissues of *Citrus* are mainly obtained after callus is initially induced to form adventitious embryos from which plantlets develop. *Citrus* is the most important genus where in vitro nucellar polyembryony (a form of somatic embryogenesis) is carried out. It has now become evident that nucellar polyembryony is not only possible in those species that naturally exhibit polyembryony, but also in mono-embryonic species that only form one embryo under natural conditons. However the mono-embryonic species of *Citrus* such as the Shamouti orange react less well than the naturally polyembryonic species (Sharp et al., 1982). The in vitro induction of polyembryony in *Citrus* is possible with fertilized or unfertilized ovules, and the embryogenesis can take place directly or indirectly. Embryo formation in *Citrus* is promoted by NAA, kinetin and complicated mixtures of compounds (coconut milk, yeast extract, and malt extract). Litz et al. (1985) have written a recent review on the propagation of *Citrus* plants.

The vegetative propagation of *Citrus* by nucellar polyembryony has the following advantages (Button and Kochba, 1977; Button and Borman, 1971a, b, c):

1. The mass propagation of mono-embryonic species is made possible.
2. Virus-free plants are made available since generally viruses are not transmitted during polyembryony.
3. In vitro culture can be advantageous by the induction and selection of mutants.
4. If rejuvenation occurs as a result of the embryogenesis this often results in stronger growth. Rejuvenation is sometimes seen as a disadvantage because often many years are needed before flowering re-occurs.

Mango *(Mangifera indica)* is also capable of somatic embryogenesis starting from the nucellus or cultured polyembryonic ovules. The regeneration of somatic embryos is easy from highly polyembryonic cultivars, although also possible with mono-embryonic cultivars (Litz et al., 1985), the rate of success is not so high. Litz et al. (1985) also showed that other tropical fruit trees can demonstrate the morphogenetic potential of the nucellus from naturally polyembryonic treees.

20.6. Regeneration of plants from single cells

It was initially very difficult to induce division in single cells in vitro. Muir et al. (1954, 1958) were the first to grow single cells of tobacco successfully; they placed a single cell on a filter paper which was laid on a mother callus. This method is referred to as nurse culture in the literature since the mother callus provides the single cell with necessary hormones and nutrients through the filter paper and so brings about growth and division of the cell. Almost immediately after this discovery Bergmann (1960) developed a much better method. He filtered a suspension culture giving rise to a population of separated cells over which he had a measure of control. These cells were then mixed with a nutrient medium to which agar was added; the still warm medium (so that pouring is possible) was then poured in a thin layer into a Petri dish. Using this method Bergmann (1960) was able to bring about cell division and callus formation in approximately 20% of the cells of tobacco and bean. His method soon became very popular and is still used. In the early research the nutrient media used in single cell culture usually contained two powerful cell division stimulators: 2,4-D and coconut-milk. In later work synthetic media have been utilized.

Benbadis (1968) also succeeded in single cell culture in a droplet suspended under a microscope slide. This was not very successful and was replaced by the growth of cells in so-called micro-chambers, in which the cells are inoculated onto agar next to an inoculated mother callus which promotes cell division (Street, 1973).

The following are used as starting material for single cell culture (Thomas and Davey, 1975; Thomas et al., 1979):

1. A suspension culture which is filtered through a fine mesh sieve, nylon cloth or cheese cloth to produce a cell suspension.
2. Callus of other tissues which is separated into single cells either mechanically or with the help of the enzyme pectinase. This method is not used very much nowadays.
3. Tissues such as the lower epidermis which after treatment with enzymes produce protoplasts (cells without the cell wall). These then have to form a cell wall. The isolation and culture of protoplasts is covered in Chapter 24.
4. Pollen grains.

The fractionation of cultured cells can be achieved either by utilizing the differences in size (e.g. sieving: passing through beds of glass beads, through nylon or stainless steel screen) or the differences in specific gravity (e.g. by discontinuous density gradient centrifugation in a solution of the correct osmotic potential). Both of these methods have been recently described by Fujimura and Komamine (1984).

The plating efficiency (establishment of the single cell culture) is given by the percentage of inoculated cells which form a callus colony. This process is dependent on many factors: the genotype, cell density of the separated cells, the rate of cell division in the starting material, age of the suspension culture, the presence of the mother callus, composition of the nutrient medium etc.

Regeneration of plants from single cells is as follows. A callus colony must initially develop from the cell, from which embryos or adventitious shoots are formed. More details are given in Section 20.5.4. Some important plants from which complete plants have been regenerated from a single cell are: *Nicotiana tabacum, Daucus carota, Cichorium endivia, Asparagus officinalis, Brassica napus* (Kartha et al., 1974) and *Citrus sinensis* (Button and Kochba, 1977). When regeneration of a plant from a single cell in unsuccessful, it is usually due to one of the following (Thomas et al., 1979):

1. The cells are insufficiently de-differentiated and rejuvenated.
2. The cells are not totipotent, usually genetically determined.
3. The factors to induce embryogenesis and organ formation are not known.

20.7. Synthetic seeds

To make the production of somatic embryos practically applicable a delivery system must be available by which the embryos survive 'sowing', develop further (for which nutrients are required) and finally germinate. To achieve this the embryos formed in vitro are coated and then known as synthetic or artificial seeds (Demarly, 1986; Redenbaugh et al., 1986; Gray, 1981, 1986).

Two systems have been developed for automated handling of these synthetic seeds:

1. By encapsulation (Redenbaugh et al., 1986). A gel, containing nutrients is coated around the somatic embryo. An encapsulation machine has already been developed which can produce and sort 6,000 single embryo capsules per hour.
2. Sowing the somatic embryos by fluid drilling (Gray, 1981, 1986). This process has been developed by seeds firms who wanted pre-germinated seeds to achieve homogeneous germination in the field. The embryos are also packed in a gel containing additives before being sown in soil.

The production of 'seeds' by coating somatic embryos and obtaining plants from these encapsulated embryos is far from easy because (Demarly, 1986; Redenbaugh et al., 1986; Gray, 1986):

1. The development of artificial seeds that are stable for several months requires chemical and physical procedures for making the embryo quiescent.
2. The artificial seed needs to be protected against desiccation when stored under dry conditions. If drying-out takes place, then the seed may enter a dormant phase. The embryos also need to be protected and stabilized during transport, storage and sowing (as is the case for normal seeds).
3. Recovery of plants from encapsulated somatic embryos is often very low; this is due to:

- Incomplete embryo formation.
- Difficulties to arrest growth.
- Difficulties in creating an artificial endosperm within the capsule (the embryo must be provided with nutrients). Hardening treatments (pretreatment of embryogenic suspension cultures: 12% sucrose, high inoculum density, chilling, ABA) to increase survival of encapsulated carrot embryos were described by Kitto and Janic (1985, 1985a).

4. The embryo must be protected (against e.g. micro-organisms) by the use of antibiotics and fungicides. However, at a later stage micro-organisms may be advantageous (mycorrhizas).

The problems encountered with the production and application of synthetic seeds are far from being solved. They are bound to be more expensive than ordinary seeds, although this initial price difference may be compensated (Demarly, 1986):

1. By higher output due to the selection of super genotypes.
2. Because male and female parental strains no longer need to be maintained.
3. Because research into male sterility may be unnecessary.
4. Because 'seed production' becomes possible in plant species where no (real) seeds are formed.
5. Because plant breeding is changing dramatically (e.g. by the application of genetic manipulation).

21. Somaclonal variation

The greatest disadvantage of some in vitro methods of vegetative propagation (adventitious shoot formation, callus culture, cell suspension culture, protoplast culture), is the chance that genetic variation and mutations may occur. Genetic variation is used to describe heritable variation that is sexually transmitted to the progeny of plants regenerated from cultured cells; the term mutant is reserved for the special case in which a trait is transmitted meiotically according to well-established laws of inheritance; the term variant is used when the nature of the heritable change(s) is not known (Meins, 1983). When genetic variation is observed in plants which have been regenerated in vitro (often the result of genetic instability), then the term somaclonal variation is used (Scowcroft, 1985). However, it should be remembered that variation and mutations can also occur during in vivo propagation (Orton, 1984). In the case of *Saintpaulia ionantha* and *Begonia* × *hiemalis* (a triploid) mutations may arise due to adventitious shoot formation whether in vivo or in vitro. If cytokinin is added, as is usually the case in vitro, than the chances of mutation are increased (Scowcroft, 1985; Ando et al., 1986).

The occurrence of mutations is not always easy to ascertain since:

1. The expression of mutations is often strongly dependent on the environmental conditions.
2. Micro-mutations often occur which are difficult to identify.
3. Cytological methods are often needed for the identification of the mutants, which call for time consuming specialized techniques. Cytological techniques alone may not be adequate for the identification of some micro-mutations and techniques such as electrophoresis to identify iso-enzyme patterns may have to be used.

As well as somaclonal variation epigenetic variation can also take place (as a result of a change in the gene expression), which is reversible and not hereditary. A few examples of epigenetic variation are:

1. With in vitro propagation the chance exists that virus-free plants arise either during meristem culture or adventitious shoot formation. These virus-free plants may differ completely in their external appearance from plants possessing the virus.
2. Different phenotypes can occur as a result of the use of cytokinin or auxin (a high degree of branching, fasciation, thickening, etc.), which although not mutations have the appearance of being so.
3. Plants may have an abnormal appearance as a result of the physiological disease called vitrification.
4. Plants can undergo changes (such as in the number or morphology of stomata, or in the composition of the cuticular wax layer), due to the special in vitro climate (e.g. a very high relative humidity).
5. Consideration must be given to the fact that after regeneration in vitro, morphogenetic phenotypes can arise, which have the appearance of mutants with respect to the starting material: depending on the original position in the mother plant, an explant can regenerate vegetative or generative shoots (Pierik, 1967), sometimes a shoot develops with adult and sometimes with juvenile characteristics (Beauchesne, 1982).

In recent years it has been agreed by both pure and applied research workers that the following factors determine the chance of mutation and the mutation frequency during in vitro culture:

1. The method of vegetative propagation used (Fig. 21.1).
2. If the plant used is a chimera.
3. The type of regulator in use.
4. The type of tissue used.
5. The starting material.
6. The number of times subcultured.

These points are dealt with separately below.

The method of vegetative propagation
Generally speaking it can be said that the more the organizational structure of a plant is broken down, the greater is the chance of mutation occurring. Genetic stability is almost certainly preserved if the in vitro propagation is carried out using the single-node or the axillary-bud method. If adventitious shoot formation occurs as a result of the use of regulators then the chances of mutations occurring are increased.

Whether mutations occur as a result of adventitious shoot formation is dependent on the following factors:

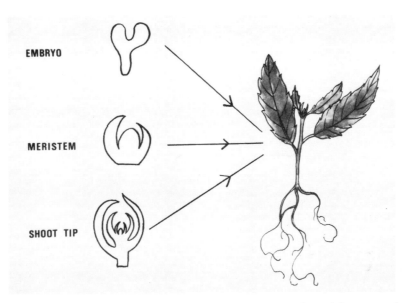

Fig. 21.1. The genetic stability of a plant is preserved when it is formed from an embryo, meristem, shoot tip or an axillary bud, without adventitious shoot formation or somatic embryogenesis.

1. The method of production of the adventitious shoots. If a shoot arises from a single cell than the chances of mutation are greater than if it arises from more cells where only sporadic mutations occur. This has been demonstrated with onion (Hussey and Falavigna, 1980) and *Saintpaulia* (Norris et al., 1983).

2. If a chimera is used, then often separation of the chimeric characteristics takes place, due to the fact that the adventitious shoot usually develops from a single cell. A few classic examples of this are: sugar cane (D'Amato, 1978; Tangley, 1986), *Yucca elephantipes* cv. *variëgata* (Pierik and Steegmans, 1983a), *Chrysanthemum morifolium* (Earle and Langhans, 1947a, b; Bush et al., 1976; Poisson and Maia, 1974), the Sim-types of *Dianthus caryophyllus* (Hackett and Anderson, 1967; Dommergues and Gillot, 1973), *Neoregelia carolinae 'Flandria'* (Fig. 21.2), *Ananas comosus 'variëgatus'* (Pierik, Steegmans and Sprenkels, unpublished data) and *Begonia* 'Aphrodite Radiant' (Hakkaart and Versluijs, 1984). George and Sherrington (1984) also listed the following plants which easily loose their chimeric characteristics in vitro: *Crypthanthus tricolor, Dracaena marginata tricolor, Dracaena warneckii, Cordyline terminalis* 'Celestine Queen' and *Hosta decorata*.

3. There are indications that there is a greater chance of somoclonal variation especially with *Gramineae* when adventitious shoots arise in

233

Fig. 21.2. In vitro vegetative propagation of *Neoregelia carolinea* 'Flandria'. Left: the experimental chimeric plant. Right: a plant, which arose after adventitious shoot formation in vitro, from which the chimeric characteristics have been lost. It is no longer three colours but has become homogeneous green (Pierik and Steegmans, unpublished results).

callus culture, while there is a much smaller chance in the case of somatic embryogenesis from callus cultures (Vasil, 1986).

4. The genotype. Some plants seldom show the tendency to mutate after adventitious shoot formation: many bulbous plants (lily, hyacinth, *Ornithogalum*, etc.) and most *Saintpaulia* cultivars. However, with other plants such as *Freesia* hybrids (Fig. 21.3, Pierik, unpublished data) and *Begonia* 'Aphrodite Radiant' (Hakkaart and Versluijs, 1984) the mutation rate is high. In addition certain cultivars of some plants (oat, potato, *Begonia* × *hiemalis*) very easily mutate (Scowcroft, 1985; Evans and Bravo, 1986). With the formation of adventitious shoots it can also be seen that tetraploids may regenerate e.g. in *Solanum* hybrids (Hermsen et al., 1981), *Nicotiana tabacum* 'Samson' (Dulieu, 1972) and *Kalanchoë blossfeldiana* (Pierik and Karper, unpublished data). Somaclonal variation has been extensively studied in the following important plant species: corn, tomato, potato and sugar cane (Scowcroft, 1985).

Mutations often appear during callus, cell and protoplast cultures (Bayliss, 1980; D'Amato, 1978). It is striking that callus cultures are genetically extremely stable in some cases such as with oil palm (Jones, 1983) and *Anthurium andreanum* (Pierik and Steegmans, unpublished data).

234

Fig. 21.3. Left: the inflorescence of a mutated *Freesia* hybr., obtained after adventitious shoot formation in a callus produced from a flowerbud of the *Freesia* depicted on the right (van der Meys, unpublished results).

Mutations seldom occur with *Anthurium andreanum*, even after treatment with X rays. Moreover it is noteworthy that with the callus culture of *Daucus carota, Oryza sativa* and *Triticum aestivum* polyploids and aneuploids arise, although normal diploids may be produced after regeneration (D'Amato, 1978); this shows that with these plants regeneration can selectively take place from cells which are normal.

An extremely striking example of genetic instability is seen in the potato cultivar 'Russet Burbank'; plants which have originated from leaf protoplasts of this cultivar are virtually always mutated (Shepard et al., 1980). The same results were found with the potato cultivar Bintje (Ramula et al., 1984). Genetic instability has also been established in *Nicotiana tabacum*, when plants are regenerated from protoplasts derived from cotyledon tissue (Dulieu and Barbier, 1982).

Regulators
In the literature it has been reported that the use of some regulators such as 2,4-D, NAA and synthetic cytokinins lead to a (large) increase in the number of mutations. Kinetin appears to induce selectively cell division in cells which are naturally (in vivo) polyploid (D'Amato, 1977). The use of high concentrations of BA strongly induced somaclonal variation in tobacco (Evans and Sharp, 1986).

235

Type of tissues and starting material used
The chance of mutations arising is less when the starting material is undifferentiated tissue (pericycle, procambium and cambium) rather than differentiated tissue such as pith. Cell differentiation in vivo can result in the production of polyploid cells (D'Amato, 1977).

Moreover, the starting material has an important effect on the production of mutations. Polyploids species exhibit chromosomal abnormalities more often than diploids (Scowcroft, 1985).

Number of subcultures
Repeated subculturing in vitro increases the chance of mutations; this is true for protocorm culture of orchids, for axillary and adventitious shoot formation and especially for callus-, suspension-, and single-cell cultures. For this reason, in practice, repeated subculture is avoided in vitro. The question as to whether the fluted leaf variant occurring in multiple recycled leaf cultures of *Saintpaulia* is genetic or epigenetic in origin has yet to be satisfactorily answered (Cassells and Plunkett, 1986).

How far the loss of regenerative potential in vitro, especially by repeated subculturing, is a result of mutation is very difficult to determine. Sometimes there is a marked correlation between the loss of regenerative potential and the production of mutations; with *Daucus carota* is has been shown that suspension cultures remain diploid, do not deviate in the number of chromosomes and remain totipotent, while callus cultures are aneuploid and are also not regenerative (Thomas and Davey, 1975). However, in many other cases there is no correlation between the loss of the regenerative potential and the production of mutations; for example, tobacco retains a high regenerative potential despite genetic instability in vitro.

In conclusion it can be said that relatively little is known about the cause of genetic instability, especially with callus-, suspension-, cell- and protoplast-cultures. It is obvious that the induction of cell division in differentiated tissues resulting in callus cultures strongly increases the chances of mutations. D'Amato has concluded the following concerning cytogenetic variability in callus culture (cf. Reuther, 1985):

1. Nuclear conditions in vivo.
    1.1. Occurrence of polysomaty due to endoreduplication.
    1.2. Aneusomaty, a rare cause of variation in chromosome number.
    1.3. Naturally occurring mosaics and chimeras.
    1.4. Structural changes in the chromosomes and other genetic changes.

236

2. Nuclear processes at the time of callus induction.
   2.1. Mitosis in diploid and endoreduplicated cells.
   2.2. Endoreduplication followed by mitosis.
   2.3. Nuclear fragmentation (amitosis) followed by mitosis.
   2.4. Differential DNA replication (amplification).
3. Nuclear conditions in callus and suspension cultures.
   3.1. Diploidy and stability at the diploid level.
   3.2. Polyploidy, both pre-existing and originating in callus.
   3.3. Aneuploidy and selective advantage of particular karyotypes.
   3.4. Chromosomal and gene mutations, pre-existing rather than originating in culture.
   3.5. Haploidy. Haploid cells and tissues appear to be susceptible to chromosome doubling.

Genetic instability may be caused by the following (Bayliss, 1980; D'Amato, 1978; Sunderland, 1973; Scowcorft, 1985):

1. Polyploid cells can arise naturally (in vivo) in plants, possibly due to endomitosis and/or nuclear fusion. This endopolyploidy, which has been shown in pith tissue of tobacco, as well as pea, can lead in vitro to a mixture of diploid and polyploid cells. If selective cell division takes place in polyploid cells, then these become dominate over the diploid cells during successive subculturing. It must be borne in mind that polyploidy also occurs in callus cultures of plants such as *Crepis capillaris, Daucus carota and Helianthus annuus,* in which no polyploid cells arise in vivo. In other words: polyploidy can also arise in vitro. Well known examples of this are *Pelargonium zonale, Nicotiana tabacum, Nicotiana alatum, Lycopersicon lycopersicum* and *Medicago sativa* (Evans et al., 1984).
2. By using auxin and cytokinin polyploid cells (see 1) can be selectively induced to divide, as has been demonstrated for pea. On the other hand it appears that synthetic auxins like NAA and 2,4-D which mainly result in non-organized growth of cells can give rise to abnormalities during cell division.
3. When aneuploid cells arise (e.g. with *Nicotiana tabacum* and sugarcane hybrids) it is assumed that these are a result of in vitro culture, since very few aneuploids arise in vivo. The auxins NAA and 2,4-D are often assumed to be responsible for the occurrence of aneuploidy in vitro, although this has yet to be authenticated.
4. An explanation of the formation of haploids, triploids, pentaploids, etc. can be found in D'Amato (1977).

Somaclonal variation is particularly interesting in plants which natural-ly show little variation, or in which variation is difficult or impossible to induce. The frequency of aneuploids and polyploids should be kept as low as possible if somaclonal variation is to be utilized as a means of variation (Hermsen, personal communication). So far no important gen-etic improvement has been obtained from tissue culture, in any crop of economical importance using somaclonal variants (Vasil, 1986). Plant breeders already 'use' somaclonal variation in a few plants species (sugar cane, tomato, potato); Orton (1984) gave a list of commercially impor-tant plants in which research has indicated the possible use of somaclonal variation for crop improvement. Vilaplana and Mullins (1986) establish-ing grape-vines from somatic embryos, showed that a number of grape-vines exhibited somaclonal variation of potential interest in viticulture, but the low yields of the plants have so far limited it's use.

Further information on somaclonal variation can be found in the fol-lowing review articles: Flick and Evans (1983a); Larkin and Scowcroft (1981, 1985); Maliga (1984); Orton (1983, 1984); Thomas (1984); D'Amato (1977, 1978, 1985); Bayliss (1980); Earle and Demarly (1982); Evans et al. (1983); Flick et al. (1983); George and Sherrington (1984); Reisch (1983); Mestre and Benbadis (1985); Cassells (1985); Evans and Sharp (1986).

In recent years researchers have directed increasing effort towards the isolation of mutants from plant tissue cultures (without applying any selection) as a result of somaclonal variation. Daub (1986) lists a number of studies where plants resistant to a particular disease were obtained from cultured cells in the absence of selection. For further details on the isolation of mutants from cultures preselected with toxins or pathogens the reader is referred to Section 25.2.4.

22. Test tube fertilization

The task of the plant breeder can be made difficult by any of the following eventualities: the pollen fails to germinate on the stigma, the growth of the pollen tube in the style partially or completely stagnates, no fertilization takes place, the fertilized egg cell does not develop in vivo and aborts, or abscission of the ovaries occurs prematurely (Rao, 1965; Rao and Rangaswamy, 1972). If no fertilization takes place after self pollination or cross pollination then this is referred to as self incompatibility or cross incompatibility. In some cases the plant breeder must resort to special procedures to bring about fertilization e.g. by ovule fertilization (here the pollen is artificially brought into contact with the ovules).

It was not until 1962 that did the idea arose of bringing about fertilization in vitro when this was not possible in vivo. Despite the fact that little research had been carried out in this area a few interesting examples of 'test tube fertilization' were found. A brief review of this work is given below based on the following review articles: Rao, 1965; Rao and Rangaswamy, 1972; Rangaswamy, 1977; Yeung et al., 1981; Zenkteler, 1980; Collins et al., 1984. A comprehensive list of plant species and genera in which in vitro fertilization is carried out is given in Pierik (1979), Collins et al. (1984), and Bhojwani et al. (1986).

In vitro fertilization is of particular importance if the incompatibility is present on the stigma or in the style.

In vitro fertilization can take place in three different ways:

1. Stigma fertilization (Fig. 22.1). In this method an emasculated flower is externally sterilized and then isolated in vitro. Pollen from a ripe anther (which has been externally sterilized) is then placed on the stigma. This method, which is similar to fertilization in vivo, can be used, if for example, the ovaries fall off the plant prematurely, resulting in a lack of progeny. Using stigma fertilization success has been achieved with: *Nicotiana rustica, Nicotiana tabacum, Petunia viola-*

239

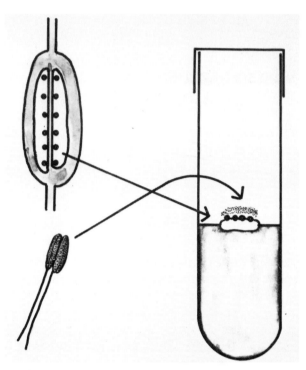

Fig. 22.1. Schematic illustration of stigma pollination in vitro. For further explanation see text.

cea, Antirrhinum majus, Pisum sativum, Lathyrus odoratus, Zea mays (Zenkteler, 1980; Pierik, 1979; Sladky and Havel, 1976) and *Glycine* species (Silvoy et al., 1984).

2. Placental fertilization (Fig. 22.2). An intact flower is externally sterilized and placenta explants with unfertilized ovules are dissected under a stereomicroscope, and inoculated onto a nutrient medium. At the same time anthers which are still closed and at a stage where they would be just about to open in vivo are externally sterilized. The anthers are opened under sterile conditions and the pollen grains placed near the ovules. After this, time is required to determine whether the pollen grains germinate, if they penetrate the embryo sac and whether fertilization follows. Placental fertilization is practiced with members of the *Caryophyllaceae*, with *Gossypium* (Refaat et al., 1984), and *Zea mays* (Truong-Andre and Demarly, 1984).

3. Fertilization of an isolated ovule without a placenta. This method is the same as in 2 from the time that the ovule is isolated in vitro. There has been little success with this method since it is extremely difficult to induce embryo formation in in vitro fertilized ovules. This

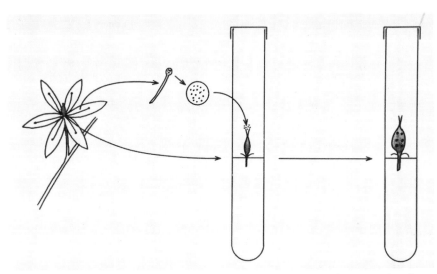

Fig. 22.2. Schematic illustration of placental pollination in vitro. For further explanation see text.

agrees with the information on embryo culture given in Chapter 16. For this reason placental fertilization is virtually always used.

In vitro fertilization can be used in the following cases:

1. Placental pollination is sometimes possible when the plants are completely self incompatible in vivo. Examples of this are: Petunia axillaris (Rangaswamy and Shivanna, 1967) and *Petunia* hybrids (Niimi, 1970).
2. Cross fertilization may be possible in vitro even if it is impossible in vivo. Iwai and Kishi (1986) were able to produce hybrid plants after test tube fertilization of ovules of *Nicotiana alata* with pollen from *Nicotiana tabacum*. Intergeneric crosses can also be achieved in vitro, as seen with different members of the *Caryophyllaceae*; for this family it has been shown that pollen grains germinate better with placental fertilization in vitro than on the stigma in vivo.
3. Production of haploids by parthenogenesis. Hess and Wagner (1974) obtained a haploid *Mimulus luteus* by pollination with pollen grains of *Torenia fournieri*; by prickle pollination unfertilized egg cells of *Mimulus* developed. It is not yet known whether this method is applicable to other species.
4. The abscission of a flower or ovary is sometimes unavoidable. In such a case stigma fertilization may be effective.
5. To study the physiology of the fertilization.

241

In general little is known about the conditions necessary for fertilization in vitro. However, it seems certain that:

1. The pollen grains and the ovules must be in the correct physiological and morphological state.
2. The choice of nutrient medium is extremely important. It is not surprising that this choice is very difficult since different processes have to take place one after the other: germination of the pollen grains, fertilization, and growth of the embryo into a seed. A complex mixture of compounds is often used to induce growth of the embryo.
3. When sterilizing flowers for use with the stigma fertilization, care should be taken that the stigma is not in contact with the sterilizing agent for too long or the exudate on the stigma will be dissolved.
4. With stigma fertilization it is better not to remove the sepals from the flower, since they encourage the growth of the ovary.
5. Rao (1965) and Rao and Rangaswamy (1972) showed with *Nicotiana rustica* that stigma fertilization may still be possible despite failure of placental fertilization.
6. Temperature may be a decisive factor. Balatkova et al. (1977) found that the fertilization of *Narcissus* in vitro required a low temperature, and never took place at 25 °C. However, with *Papaver somniferum* it appears that low temperatures are unfavourable.

23. The production of haploids

23.1. General introduction

With sexual (generative) reproduction the number of chromosomes, reduced to a half as a result of meiosis, are doubled again by fertilization (fusion of male and female gametes). If meiosis takes place in a diploid plant (2 sets of chromosomes denoted by $2n = 2x$) cells arise with $n = x$ chromosomes; if such a cell gives rise to a plant without fertilization this is termed a monohaploid plant, which only has one set of chromosomes (x). Similarly with a tetraploid plant (4 sets of chromosomes, denoted by $2n = 4x$) meiosis gives rise to cells with $2n = 2x$ chromosomes. If plants are formed from such cells without fertilization, they are called dihaploids. Generally speaking, haploid cells or individuals are those in which the original chromosome number has been reduced by half.

If the number of chromosomes of a monoploid plant ($n = x$) is doubled, either spontaneously or by induction (with the help of colchicine) a homozygous diploid individual is formed with 2 identical sets of chromosomes ($2n = 2x$): this doubling (duplication) is very important since monoploids are sterile. If, as a result of fertilization, fusion of two cells occurs each with $n = x$, but with non-identical sets of chromosomes, then a heterozygous diploid individual is formed. However, homozygous individuals give rise to identical sex cells (gametes) by meiosis. If two homozygotes, which are strongly different genetically, are crossed with each other, then the progeny will be completely identical but heterozygous. The latter point is of vital importance to plant breeders and growers as we have already established in Chapter 20 (vegetative propagation as a means of producing homogeneous progeny).

Obviously plant breeders have searched intensively for haploids that have arisen spontaneously or attempted to make them artificially. The number of plant species known at the present time, in which haploids have arisen spontaneously in vivo is more than 100, but as a rule the frequency is very low. The most important source of finding these haploids is 'twin seeds' (a seed with two embryos). In a twin seed an n-n twin may be found (as the result of the growth of one unfertilized egg and one synergid) or a 2n-n twin (arising from one fertilized and one unfertilized egg-cell).

The production and occurrence frequency of haploids in vivo can be improved by many different techniques (Hermsen, 1977, 1984; Hermsen and Ramanna, 1981; Sneep, 1983; Wenzel, 1980).

1. Gynogenesis. This is the development of an unfertilized egg-cell, either as a result of delayed (retarded) pollination, through use of abortive pollen (pre-exposed to ionizing irradiation) or through the use of alien pollen (sometimes even pollen from the same species). Gynogenesis is found especially in interspecific crosses e.g. the cross between *Solanum tuberosum* and *Solanum phureja* whereby dihaploid $(2n = 2x)$ potatoes are produced, which were originally tetraploid $(2n = 4x)$. With gynogenesis it is vital that viable endosperm arises, otherwise the embryo aborts in vivo.
2. Androgenesis. In this case the nucleus of the egg cell is eliminated or inactivated before fertilization; the haploid individual being produced by the development of the egg-cell containing the male nucleus.
3. Genome elimination which arises as a result of certain (intergeneric and interspecific) crosses. With such a cross, fertilization occurs, but soon afterwards one genome is eliminated giving rise to an embryo that only has one genome present, and is consequently haploid. An important example of this is the interspecific cross between *Hordeum vulgare* and *H. bulbosum*, which results in the production of a haploid *H. vulgare*. This will be referred to again later.
4. Semigamy. In this case the nucleus of the egg-cell and the generative nucleus of the germinated pollen grain divide independently, resulting in a haploid chimera.
5. Chemical treatment: toluene blue, maleic hydrazide, nitrous oxide and colchicine. By the addition of chloramphenicol and para-fluorphenylalanine chromosome elimination can be induced.
6. Shocks with high or low temperatures.
7. Irradiation with X-rays or UV light.

The main topic in this Chapter is the induction of haploids in vitro: the stimulation of reduced female gametes (gynogenesis) or male gametes (androgenesis) to autonomously develop into haploid individuals (Hermsen, 1977). In gynogenesis, which mainly takes place in vivo, the female egg-cell or synergid is stimulated to grow without being fertilized.

Fig. 23.1. Different stages in the development of haploids from isolated anthers of *Nicotiana tabacum.* From left to right: anthers just isolated; plantlets developed after 6 weeks in darkness; after 6 weeks darkness and 1 week in the light; after 6 weeks darkness and 3 weeks in the light. Growing temperature 25 °C.

In androgenesis, which virtually only takes place in vitro, the vegetative or generative nucleus of a pollen grain is stimulated to develop into a haploid individual without undergoing fertilization.

In the literature concerning the induction of androgenesis in vitro (Pierik, 1979; Bhojwani et al., 1986) it is clear that the members of the Solanaceae are especially responsive and are capable of regenerating haploids from isolated anthers (Fig. 23.1) or individual pollen grains. In the families of the *Cruciferae, Gramineae* and *Ranunculaceae* different genera can also form haploids (Bajaj, 1983; Rashid, 1983); striking results have also been obtained in the last few years with the anther culture of rice (Reeves, 1986). A remarkable correlation can be seen between the regenerative potential that is found in vegetative propagation (Chapter 20) and that of anthers. Maheshwari et al. (1983) reported that haploid induction is possible especially in the Angiosperms: 247 species and hybrids, from 88 genera and 34 families are able to produce haploids in vitro. There is sometimes a striking difference in haploid induction even within a single genus, from very easy to impossible. There are typical differences within a genus in *Oryza* (rice), *Lycopersicon* (tomato), *Arabidopsis, Solanum, Triticum* and *Nicotiana.* There has so far been very little success with trees and shrubs (Mott, 1981). More information concerning haploid induction in trees is given by Radojevic and Kovoor (1986).

245

It is perhaps interesting that haploid induction was discovered late in the development of tissue culture. Guha and Maheshwari (1964, 1966) working in India first showed that in vitro isolated anthers of *Datura innoxia* were able to form haploid embryos. Bourgin and Nitsch (1967) obtained the first haploid plants from isolated anthers of *Nicotiana*. Since 1964 there have been many publications concerning the production of haploids in vitro; this points to it's importance to plant breeders and geneticists. The advantages of haploids are surveyed below (Sunderland, 1973; Hermsen, 1977, 1984; Hermsen and Ramanna, 1981; Jensen, 1977; Reinert and Bajaj, 1977a; Debergh, 1974, 1978; Wenzel, 1980; Raghavan, 1986):

1. As a result of haploid induction followed by chromosome doubling, homozygosity is achieved in the quickest possible way making genetic and breeding research much easier. The genetic segregation is simplified in homozygotes, recessive genes not being masked by a dominant ones. The production of homozygotes is especially important for the obligate cross breeders, to which dioecious and selfincompatible plants belong. With cross pollination the heterozygosity is enhanced. If self pollination is also possible with cross breeders then they are very difficult to obtain as homozygotes due to inbreeding. Rapid attainment of homozygosity is also desirable with self pollinators, although self pollination as a rule results in a high rate of homozygosity. However, it can still take 5–7 generations of self pollination before complete homozygosity is obtained.
2. Homozygosity is still more important for those plants which have a very long juvenile phase (period from seed to flowering), such as fruit trees, bulbous plants and forestry trees. Even if repeated self pollination is possible achievement of homozygosity, especially in this group of plants, is an extremely long process.
3. As a result of homozygosity lines become available which make the production of pure F_1-hybrids possible; a F_1-population should, in principle, be as homogeneous as possible.
4. When working with plants that are normally polyploid, it is very useful by haploid induction to work at a lower ploidy level; studying the inheritance and combining desirable characteristics is much easier at a diploid level rather than in a tetraploid, which means that this technique is particularly efficient for autotetraploids.
5. Monohaploids have the advantage for the mutation breeder that the recessive mutations (A→a) are immediately discernable. This is not the case with a diploid which mutates from AA to Aa; self pollination of Aa gives rise to only one in four of the double recessive mutant. If

a large population of haploid cells are available, then mutation induction and selection is possible on a large scale; so it is for example conceivable that resistance against phytotoxins could be selected for in a haploid cell population. An added advantage of mutation induction in haploid cells is that chimeras are avoided. Examples of haploid mutation induction are given by Nitsch et al. (1969a), Sunderland (1973a), and Bajaj (1983).

6. By haploid induction followed by chromosome doubling it is possible to obtain exclusively male plants. An important example of this is *Asparagus officinalis*, where male plants have a higher productivity and yield earlier in the season than female plants. Female *Asparagus* plants are designated XX and males XY; by crossing XX and XY 50% female and 50% male plants are obtained. If haploids are produced from anthers from male asparagus plants these are either X or Y; chromosome doubling of Y results in super male plants YY which can subsequently be vegetatively propagated. If XX is crossed with YY then only males XY plants result (Hondelmann and Wilberg, 1973; Thévenin and Doré, 1976).

7. When homozygotes are artificially made into diploids in vivo with the use of colchicine, problems can arise which are not usually found in vitro. In vitro haploids often double spontaneously giving rise directly to homozygotes, and in addition application of colchicine is much easier in vitro than in vivo.

8. It is much better to work with haploid protoplasts rather than diploid ones for somatic hybridization. Fusion of two haploid protoplasts results in a diploid, while the fusion of two diploids results in a tetraploid.

It should be pointed out that obtaining a monohaploid in itself, is not important to plant breeders since this will be sterile. Firstly it must be doubled to attain fertility. There are two methods in vitro to double the number of chromosomes of a haploid. Spontaneously, by endomitosis, often by the formation of an adventitious embryo or shoot; or by treatment with colchicine. If homozygotes are obtained after chromosome doubling then these are not the ultimate goal of the plant breeder, since they are only starting material for further crosses which will lead qualitatively and quantitatively to the production of valuable hybrids.

23.2. Obtaining haploids in vitro

The following starting materials for obtaining haploids are described in the literature (Debergh, 1978; Sunderland, 1978, 1980):

1. Anthers. By far the most research has been carried out on isolated anthers (Fig. 23.2) which have been isolated onto solid or in liquid nutrient media. Sometimes closed flower buds are sterilized, and in other cases the still closed anthers are sterilized which have been obtained from a flower that has already opened.
2. Individual pollen grains. This is a little used technique (due to technical problems) and is dealt with briefly later.
3. Inflorescences which are usually grown in a liquid medium. This method, which was initially used for *Hordeum*, is especially useful with grasses and other plant species which have small flowers or flowers with a reduced perianth.
4. Embryo culture. If, for example, *Hordeum vulgare* is crossed with *H. bulbosum*, then fertilization takes place, although soon afterwards the chromosomes of *H. bulbosum* are eliminated. A haploid embryo of *H. vulgare* is produced although no endosperm is developed; without embryo culture this embryo would abort (Jensen, 1977).

   Another example analogous to that above is the crossing between *Triticum aestivum* and *H. bulbosum*, which results in the production of a haploid *Triticum aestivum* embryo; in this case embryo culture is also necessary (Barclay, 1975; Zenkteler and Straub, 1979).

   Kott and Kasha (1985) have published a comprehensive review article on haploid plant production and embryo culture.
5. Pseudo-fertilization. Hess and Wagner (1974) pollinated *Mimulus luteus* in vitro with alien pollen of *Torenia fournieri*; by prickle pollination a haploid cell of *Mimulus* developed into a haploid plant.

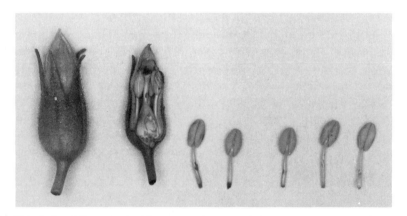

Fig. 23.2. A closed flower bud of *Nicotiana tabacum* (left) is initially externally sterilized, and then cut open in the laminar air-flow cabinet (2nd from left). The stamens (right) can then be removed. The anther filaments are removed, to prevent these regenerating plants, and the isolated anther is inoculated into the nutrient medium.

6. The development of unfertilized ovules without pseudo-pollination. The main advantage of ovule culture appears to be that the majority of regenerants are green and are more genetically stable (Dunwell, 1985). San Noeum (1976) had the first success with *Hordeum vulgare* in obtaining haploids from cells of the embryo sac, which were made diploid using colchicine. Haploids have also been produced from cultured ovules of 3 other cereals (wheat, rice, and maize). Preil et al. (1977) using the capitulum method of Pierik et al. (1973) with gerbera obtained a yellow flowering haploid plant, perhaps due to the development of an unfertilized ovule. Later, haploids have also been produced from cultured ovules of gerbera by Meynat and Sibi (1984). Verna and Collins (1984) obtained haploid *Petunia axillaris* plants from the in vitro isolated ovules with a piece of placenta attached. Keller (1986) reported that ovule culture for haploid production may be better than anther (microspore) culture for some taxonomic groups such as the *Compositae*.

Another method of inducing haploids is the combination of gametophytic irradiation and ovary culture; in *Petunia* the in vitro ovary culture of plants pollinated with irradiated pollen is so far the most efficient technique for haploid production. The irradiated pollen acts not only in stimulating gynogenesis but also takes part in fertilization (Raquin, 1986).

In summary it can be said that most of the haploid plants which are obtained in vitro have arisen from isolated anthers or sometimes from pollen. Androgenesis has become by far the most important source of haploids in vitro. This is extremely important because pollen grains are available in considerable numbers and can separately develop into haploids. Relatively little is known about the development of (un)fertilized ovaries and ovules into haploids.

The growth of individual pollen grains, how difficult that may be technically, and the eventual regeneration from these haploids has a number of special advantages over anther culture (Debergh, 1978; Sunderland, 1978; Nitsch, 1977; Wenzel, 1980; Bajaj, 1983):

1. The chance of regeneration of diploids is drastically limited by elimination of diploid tissues (septum, anther wall, tapetum).
2. The anther no longer acts as a barrier to the transport of nutrients from the medium to the pollen grains.
3. By elimination of the anter, inhibitors (ABA and toxic substances) are no longer a problem.

4. Callus formation from the anther can be avoided, resulting in far fewer chimeras; if a callus develops from a single pollen grain then it is of the same genotype, whereas if it develops from many pollen grains chimeras may arise.
5. Direct transformation often takes place from a pollen into an embryo as a result of isolation of the pollen.
6. Pollen grains are more suitable than anthers as starting material for mutation research and genetic manipulation.
7. Embryo formation can be observed better with pollen grains than with anthers.

Despite the advantages of pollen culture (Fig. 23.3), and the extensive research carried out on this topic it is still very difficult to induce individual pollen grains to form haploids. A much more complicated nutrient medium is necessary than for anther culture, and the number of haploids is far fewer. Pollen culture is unsuccessful for most plants, but can be achieved in the following species: *Petunia hybrida, Lycopersicon lycopersicum, Brassica oleracea, Nicotiana tabacum* and *Datura innoxia.*

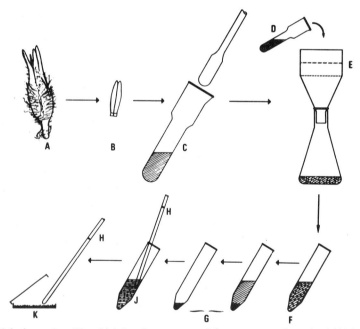

Fig. 23.3. An anther (B), which has been removed from a young flowerbud (A) is homogenized using a Potter's homogenizer (C). In this way a crude pollen suspension (D) is obtained which is then purified. The suspension is sieved (E), centrifuged (F) and decanted (G), and then inoculated by pipette (H,J) onto the culture medium in a Petri dish (K). From Debergh (1974).

Within the last few years, major advances have been made in the development of isolated microspore culture technique, particularly for *Brassica napus* (Keller, 1986).

The growth of pollen requires very specialized techniques:

1. The initial growth of the closed anthers, prior to growing the pollen grains.
2. The extraction of the pollen grains (filtering, washing, centrifuging, etc.).
3. Application of a nursing procedure.

Haploids can in principle be produced in 2 ways from isolated anthers (Debergh, 1978; Reinert and Bajaj, 1977a):

1. Directly. An embryo differentiates directly from the pollen grain (microspore).
2. Indirectly. Firstly a callus develops from the pollen grain, and then an embryo or an adventitious shoot regenerates. This type of development is not favoured since usually a mixture of ploidy levels is produced due to: the heterogeneous nature of the starting material (haploid and diploid), the addition of regulators, the spontaneous change from haploid to diploid. The haploids can often not compete with the higher ploidy level tissues.

The stage at which the anther is at the moment of isolation is of vital importance for haploid induction. Most authors (Debergh, 1978; Reinert and Bajaj, 1977a; Sunderland and Dunwell, 1977) are in agreement that the anther should be isolated when the pollen grains (microspores) inside have still not undergone their first division. Dunwell (1985) defined the correct isolation as follows: the most productive anthers are those which, when harvested, contain uninucleate microspores midway between release from the tetrad and the first pollen grain mitosis. The first cell division where a small generative nucleus and a large vegetative nucleus arise can be identified by the Feulgen-staining, to reveal the cell nuclei. In a few cases the anthers have to be isolated at a later stage, sometimes during, and sometimes after, the first division of the microspore (Nitsch, 1981; Sunderland, 1980). A wide ranging discussion of this question of when to isolate anthers and pollen grains can be found in Heberle-Bors (1985). She concluded that the developmental stage is a very complex factor, composed of at least three parts: P-grain (embryonic pollen grain) maturation, normal pollen grain maturation, and anther wall maturation.

Often the correct developmental stage can be chosen on the basis of morphological characteristics. With tobacco, flower buds are chosen in which the petals are just visible (Fig. 23.4). This method can be dangerous in that a particular morphological stage does not always correspond with the first division of the microspore. It should also be borne in mind that within any one anther not all the microspores development is synchronized; those in the centre are more advanced than those at the periphery.

With anther culture, as well as a number of haploids, chimeras and/or plants with different ploidy levels are often developed. The reasons for this last occurrence are (McComb, 1978; Thomas and Davey, 1975; Wenzel, 1980):

1. Through endomitosis (from x to 2x and from 2x to 4x) instead of haploids, diploids and tetraploids develop which are homozygous.
2. By nuclear fusion of two identical nuclei within a single haploid microspore $(x + x = 2x; 2x + x = 3x)$ diploids, triploids, etc. develop which are homozygous.
3. Regeneration of the septum and the wall of the anther gives rise to 2n individuals which are heterozygous.

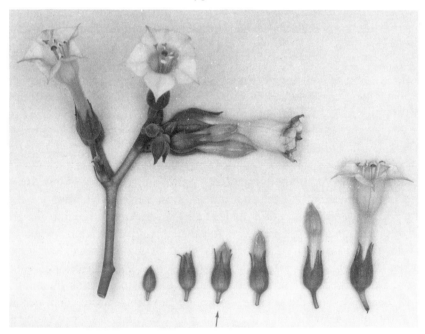

Fig. 23.4. Starting material for tobacco anther culture. Left: an inflorescence. Below: flowers in various stages of development. The anthers of the flower indicated with an arrow are often used as starting material for induction of haploids. For further explanation see text.

4. Heterogygous diploids develop by regeneration of non-reduced microspores.
5. A homozygous diploid is formed whenever there is spontaneous doubling of the haploid (Stringham, 1977).
6. Abnormalities can occur during meiosis, which lead to different ploidy levels.

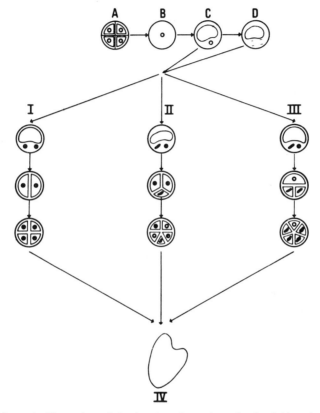

Fig. 23.5. Schematic illustration of the in vitro formation of a haploid embryo from an isolated microspore (Debergh, 1974).

A: Tetrad with 4 microspores (developed after the 1st and 2nd reduction division).

B: Isolated microspore just before the first pollen mitosis.

C: Vacuolation, resulting in the nucleus moving to the cell wall.

D: The first pollen mitosis.

I: The first pollen mitosis, resulting in the formation of 2 symmetrical nuclei which both divide.

II: First pollen mitosis resulting in the formation of 2 asymmetrical nuclei (one vegetative and one generative) of which only the vegetative one divides and the generative one aborts.

III: Note that here 2 asymmetrical nuclei are also formed, of which only the generative one divides and the vegetative one aborts.

IV: The formation of a haploid embryo.

If haploids develop from anthers (schematically shown in Fig. 23.5; Fig. 23.6), then these haploids originating from a microspore with a single nucleus can be produced in different ways (Sunderland and Dunwell, 1974; Reinert and Bajaj, 1977a):

1. In many cases after the first nuclear division (in principle resulting in a vegetative and generative nucleus) of the microspore the generative nucleus degenerates or becomes dormant and only the vegetative nucleus divides.
2. Sometimes the vegetative nucleus degenerates or becomes dormant after the first division of the microspore and only the generative nucleus divides. This unusual occurrence is seen in *Hyoscyamus* (Raghavan, 1976).
3. Two identical nuclei are formed at the first division both of which divide. This is referred to as symmetric microsporogenesis.

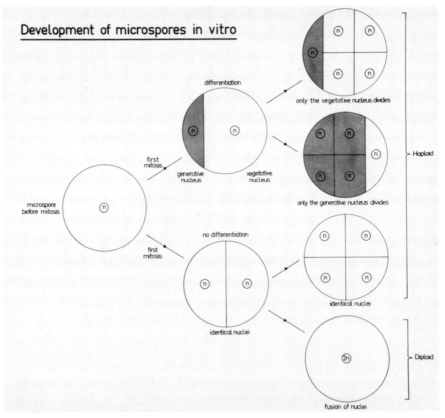

Fig. 23.6. Schematic illustration of the development of microspores in vitro. After the first mitosis non-identical (assymetrical) or identical (symmetrical) nuclei appear. The result after the second mitosis can be a haploid or a diploid. Diploid cells can arise by cell fusion.

4. In practice there is usually a mixture of all the above processes.

When producing haploids it appears that pre-treatments, particularly low temperatures, are very important. Low temperatures applied to the mother plant, a young flower bud or anther often promote embryogenesis. Sometimes high temperature may be used instead. Cold has the effect that at the first division of the microspore two identical nuclei are formed rather than one vegetative and one generative nucleus. Other pretreatments may also promote androgenesis (centrifuging the anthers, incision of the top of the inflorescence of the donor plant). It is sometimes advisable to cut the wall of the anther so that the embryos or the callus can come out more easily. Details of pre-treatments are given in: De-bergh (1978), Rashid (1983), Sunderland (1978, 1980), Nitsch (1981).

The starting material (the donor plants) and their conditioning are very essential for androgenesis. As pointed out earlier the plant family, the plant species and even the cultivar are all important. Heberle-Bors (1985) discusses the role of the genotype in haploid induction. In principle healthy strong plants should be used. The age of the donor plant, as well as the age of the flower bud both play a role. In most cereal species investigated it appears that anthers from primary tillers are more productive than those from lateral tillers; only in the case of rice there is no difference (Dunwell, 1985). In a review article (Heberle-Bors, 1985) the important question is asked as how to P-grains (embryogenic pollen grains) can be obtained depending on the pretreatment of the starting material (e.g. daylength, flower induction, low temperature, shift of sex balance towards 'femaleness' and increased pollen sterility). Heberle-Bors (1985) concluded that P-grains represent a particular form of male sterility, originating from a deviation from normal male development at the sporophytic-gametophytic transition in microspore mother cells.

Evidence has been obtained which indicates that a pollen population is dimorphous. On the one hand there is a small fraction of pollen grains, which form haploids (P-grains) in a low frequency; on the other hand there is a large fraction which does not form haploids (Rashid, 1983). For in vitro isolation obviously embryogenic anthers are selected within the population which give a high percentage of haploids (Heberle-Bors, 1985). By applying the gametocide CGA_1 Schmidt and Keller (1986) were able to screen for dimorphic pollen with a good androgenetic response.

The composition of the nutrient medium which is of such importance, can be found in the relevant specialized literature (Bhojwani et al., 1986; Pierik, 1979). A few of the most important considerations concerning the media are given below. Although solid media are often used, liquid media are becoming more popular. The sugar requirements are strongly dependent on the species and family of the experimental plant. Usually saccharose at a concentration of 2–4% is used, although higher concentrations may be required for some plants. The mineral requirements are also dependent on the particular plant being used, and a modified MS-medium or the classical Nitsch (1969) macro-salt medium are often used. The pH is generally adjusted to 5.8 before autoclaving. Regulators are usually omitted to limit the formation of callus. However, in a few cases auxin, cytokinin or a combination of both are required. It is often apparent that a complex mixture of naturally occurring plant substances are required to induce embryogenesis. Activated charcoal plays an extremely important role; it not only absorbs toxic substances (which result in a brown colour) from degenerated anthers, but also ABA which prevents embryogenesis (Johansson and Eriksson, 1982).

Growth in liquid media is often found to be very favourable. However, if liquid media are used the anthers should be floated, after the pollen grains have become free, they form embryos on the bottom of the Petri dish. It is even better to grow anthers in a thin layer of liquid medium, which is on the top of an agar plate to which activated charcoal has been added. This system is referred to as the 'double-layer method'.

It should be remembered that the composition of the nutrient medium may need to be modified for the different developmental stages during anther culture. This was demonstrated by Wenzel and Fouroughi-Wehr (1985) for potato: the following components in the media have to be changed: macro-salts, sucrose, auxin, cytokinin, activated charcoal and coconut milk.

Physical factors (such as daylength, irradiance, light quality, the light/dark cycle, day/night temperature, CO_2 feeding) during the in vitro culture can also play an important role in androgenesis (Nitsch, 1977, 1981; Bajaj, 1983; Wenzel, 1980; Debergh, 1978; Dunwell, 1985).

23.3. Problems associated with haploid induction

Many problems are encountered with haploid induction, and the percentage of anthers (pollen grains) that form viable haploid embryos is (very) small with many plants. A large number of other factors may also complicate the overall picture. The most important problems are:

1. Often no growth and development takes place in vitro, or the initial growth is followed by abortion of the embryos.

2. Diploids and tetraploids often regenerate at the same time as the haploids.
3. The regeneration of diploids etc. can, in principle, be overcome by growing isolated pollen grains. However, this technique has been developed for a restricted number of species.
4. Selective cell division must take place in the haploid microspores and not in other unwanted diploid tissues. This selective cell division is often impossible.
5. Regeneration of albinos is often difficult to avoid especially with cereals.
6. In vitro induction of haploids is often economically untenable due to the low success rate. A classic example is *Solanum tuberosum*, which is more efficiently made haploid in vivo by 'fertilizing' with *Solanum phureja*.
7. Callus formation, whether it has arisen spontaneously, or has been induced by regulators is usually detrimental.
8. There is little chance of isolating a haploid from a mixture of haploids and higher ploidy levels, since in these circumstances the haploids are easily outgrown by the stronger polyploids.
9. The selection of haploids and eventual production of homozygotes from them is time consuming since often cytogenetic research is necessary. Sometimes selection is possible by the so-called genetic (biochemical) markers.
10. The doubling of a haploid does not always result in the production of a homozygote. 'Homozygote' diploids sometimes exhibit segregation in their progeny.

24. Genetic manipulation

24.1. General introduction

Until 1970 it was only possible, with higher plants, to transfer genetic material from one individual to another by sexual means: through fusion of the egg with the generative pollen nucleus to give a fertilized egg, from which an individual can develop with characteristics derived from both the mother and the father. During sexual reproduction only specific sex cells fuse, in which the number of chromosomes has been reduced by half as a result of meiosis. The number of chromosomes after fertilization consists of half the number of the mother together with half the number of the father. The hereditary characteristics are localized in the nucleus (the chromosomes being the carriers: chromosomal genes) and the cytoplasm (within the chloroplasts and the mitochondria: cytoplasmic genes). In the case of sexual reproduction hereditary characteristics localized in the cytoplasm are usually only inherited from the mother.

A completely different way of transmitting genetic material (and so lead to a possible increase in genetic variation) takes place when two protoplasts fuse (cells, from which the cell wall has been digested leaving only the plasmalemma as a barrier); these usually originating from somatic cells. This process is called somatic hybridization (Fig. 24.1). In this case the chromosome number of the somatic hybrid is equal to the sum of that of the two 'parents' (fusion partners). In somatic hybridization in contrast to sexual fertilization the nuclear and cytoplasmic heriditary information is transmitted from both partners.

Power et al. (1970) and Power and Cocking (1971) fused somatic cells for the first time using protoplasts of maize and oat. Despite the fact that they did not obtain hybrid plants as a result of this fusion their approach to the problem was ingenious. They achieved fusion of two cells by using protoplasts which were obtained by enzymatic digestion of the cell wall.

In 1971 Takebe et al. obtained normal plants from protoplasts using

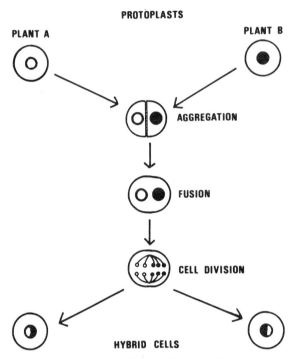

Fig. 24.1. Schematic illustration of somatic hybridization of protoplasts of 2 different plant species.

tobacco. Somatic hybridization was achieved for the first time by Carlson et al. (1972), who fused isolated protoplasts from *Nicotiana glauca,* and *Nicotiana langsdorffii* and obtained the first somatic hybrid plant.

In contrast to classical genetics, hybridization of protoplasts makes use of somatic cells rather than intact plants and their sex cells. However the term 'somatic hybridization' is not particularly apt since in vitro protoplast fusion and then hybridization can also be induced in sex cells (such as pollen protoplasts). In the last 10 years there have been spectacular scientific results and technical advances made in the field of cell hybridization, although it is still not used in practice in plant breeding (Wenzel, 1983).

The asexual transmission of genetic material appears only possible by the use of protoplasts. The cell wall is an impenetrable barrier for the fusion of cells and the transmission of e.g. organelles. Protoplasts can be obtained in large numbers from tissues, cell suspensions etc. by treatment with mixtures of enzymes which only leave the cell contents enclosed by the plasmalemma.

24.2. Description

Genetic manipulation (de Groot et al., 1982) is the molecular genetical method of transmitting genetic material from cell (protoplast) to cell (protoplast) in vitro, without the intervention of a generative phase (sexual cycle). The different types of genetic manipulation possible are given below (de Groot et al., 1982; Sneep, 1982; Schilperoort, 1983, 1984):

1. Somatic hybridization (also known as parasexual hybridization). This is the complete fusion of nuclei and cytoplasm of two protoplasts (Fig. 24.2). Sometimes there is partial transmission of genetic material, when for example one of the protoplasts is pre-irradiated (resulting in loss of chromosomes), or if spontaneous chromosomal elimination occurs.
2. Cytoplasmic hybridization. This involves the fusion of the nucleus and the cytoplasm of one protoplast (A) with the cytoplasm of anoth-

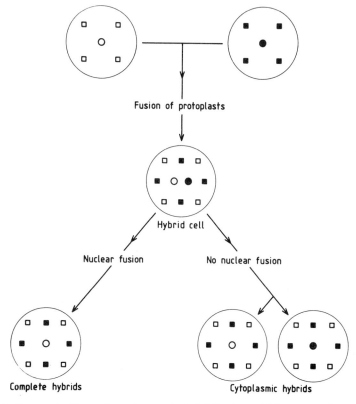

Fig. 24.2. Schematic illustration of somatic hybridization which can produce complete hybrids (left) or cytoplasmic hybrids (right).

er protoplast (B). This type of hybridization is especially interesting in connection with the transmission of e.g. cytoplasmic male sterility, which is mainly localized in the mitochondria. Cytoplasmic hybridization (Fig. 24.2), which results in so-called cytoplasmic hybrids or cybrids only occurs in a limited number of cases (Galun and Aviv, 1984; Pelletier and Chupeau, 1984).

3. Transplantation and uptake of isolated nuclei, chromosomes, chromosome fragments and organelles (plastids and mitochondria). For instance it would be interesting to implant (to build into the genome) a piece of chromosome on which the gene for disease resistance is localized. The uptake of isolated nuclei from *Vicia hajastana* cells by protoplasts of an autotrophic cell line of *Datura innoxia* has recently been achieved (Saxena et al., 1986).

4. Transformation. The addition of specific genes to a cell or protoplast by the use of recombinant DNA requires a vector (a plasmid or virus) and isolated genes (from plants, bacteria, or viruses). In this research the Ti(tumour inducing)-plasmid from the crown-gall bacteria *(Agrobacterium tumefaciens)* is often used as a vector. This plasmid leads to efficient transmission of genetic information, and is frequently used with dicotyledons, although use with monocotyledons is not completely precluded (Schilperoort, 1984). Hooykaas-van Slogteren et al. (1984) showed that two monocotyledonous plants *(Narcissus* and *Chlorophytum)* could be infected by *Agrobacterium tumefaciens.* More details on this subject are given in Section 24.9.

In points 1 and 2 the terms donor and recipient cells are not used since the fusion partners are generally equal with regard to the fusion process and integration of the genome, unlike point 3 where donors and recipients are considered viable terms. Somatic hybridization (1) has been shown to be possible for many different plants, while cytoplasmic hybridization (2), apart from work with *Nicotiana* is still very much in the initial stages of research (Galun and Aviv, 1984). The transplantation of isolated nuclei etc. (3) may well be developed in the future, but transformation (4) is already an intensive area of research (Schilperoort, 1984) and promises many possibilities (Section 24.9).

Sometimes somoclonal variation (Chapter 21), fertilization with radiated pollen grains and the induction and selection of mutants in vitro is also considered under the term genetic manipulation. As the application of genetic manipulation in the strictest sense often requires mutants for the systemic selection and isolation of hybrids this topic will be covered shortly in Section 24.8.

There are a large number of articles and handbooks (these have been consulted and used when writing this Chapter) with information on genetic manipulation, the most important of which are given below: Gleba and Sytnik (1984); Harms (1983); Potrykus et al. (1983); Linskens et al. (1983); Davey and Kumar (1983); Pelletier and Chupeau (1984); Sneep (1982); de Groot et al. (1982); Binding et al. (1981, 1982); Evans and Bravo (1983a); Ahuja (1982); Wenzel (1983); Thomas et al. (1979); Vasil (1978, 1980a); Schieder (1980); Widholm (1978, 1982); Schilperoort (1984); Cocking (1978); Thorpe (1981); Bajaj (1974); Gamborg and Wetter (1975); Dodds (1985); Potrykus et al. (1983); Shillito et al. (1986); Power and Chapman (1985).

24.3. Prerequisites for the use of genetic manipulation

Plant breeders have the following prerequisites for genetic manipulation (Sneep, 1982):

1. The hybrid cells should, in principle, be genetically stable.
2. It must be possible to regenerate plants from the protoplasts and cells.
3. The plants obtained should ideally not be female sterile.
4. The genetic information which is transmitted to another cell by cell hybridization or other techniques must be transmissible to a cultivated crop.

24.4. Somatic hybridization

24.4.1. *Outline*

To achieve somatic hybridization many phases need to be carried out and these are summarized below (Fig. 24.3):

1. Choice and growth of the plant material.
2. Sterilization of the plant material, if this has not been grown in vitro.
3. Enzymatic treatment to obtain protoplasts, and their subsequent purification.
4. Fusion, directly followed by selection of the somatic hybrids. Direct selection is only possible if the fusion products are immediately identifiable.
5. Culture of the protoplasts, cell wall regeneration and the induction of cell division. Microcalluses arise due to cell division, and then se-

lection can be carried out on these usually with the help of biochemical mutants.
6. Regeneration of plants.
7. Verification and control of the fusion products.
8. Testing the usefulness in plant breeding.

24.4.2. *Discussion of the different phases involved*

Growth, choice and sterilization of the starting material
Growth of the starting material can take place in vitro or in vivo. In vitro grown material has the advantage that sterilization is not necessary as is

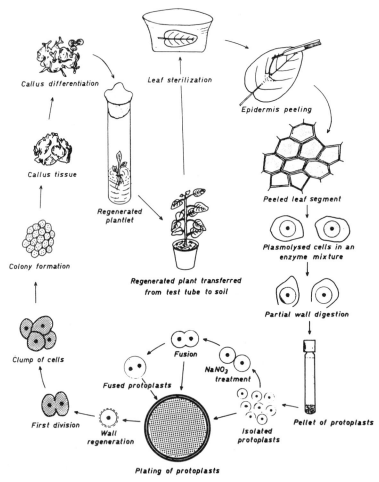

Fig. 24.3. Schematic illustration of isolation, fusion and culture of leaf protoplasts, followed by callus formation and regeneration of plants. For further explanation see text. Drawing from Bajaj (1974). Protoplasts fusion is promoted by PEG and not only by NaNO₃.

the case for in vivo grown material. When sterilizing leaves etc. extreme care should be taken as this can effect the quality of the protoplasts. Mesophyll tissue is generally used. The use of callus tissue and suspension cultures has the disadvantage that they are genetically unstable and so there is a high degree of variation. For material grown in a greenhouse or growth room the growth conditions should be optimal (Evans and Bravo, 1983a; Binding et al., 1981).

Binding et al. (1981) preferred shoot tips for making protoplasts, whereas others (see Potrykus et al., 1983) preferred seedlings, cotyledons, or hypocotyls.

The choice of the genotype can determine the success of the whole process from protoplasts to regeneration of a plant. In the case of e.g. the tomato only 1 genotype out of 14 results in a complete plant after protoplast isolation (Evans and Bravon 1983a).

Obtaining protoplasts

Protoplasts are obtained by enzymatically treating mesophyll cells, callus, cell suspensions etc., with a mixture of enzymes. The choice of enzymes, the concentration used and the duration of the treatment can be found in the specialized literature cited in Section 24.2 A mixture of cellulase, pectinase and sometimes hemicellulase is generally used, these mainly originating from microbiological sources. These enzymes dissolve cellulose, pectin (middle lamella) and hemicellulose respectively. If a cell suspension is chosen as starting material for the preparation of protoplasts then cells should be chosen which are in the exponential phase of growth, since these have thinner cell walls. During the enzyme treatment the protoplasts obtained need to be stabilized. For this reason an osmoticum is added which prevents the protoplasts from bursting. Glucose, mannitol and sorbitol are used as osmoticum. It should be borne in mind that the treatment adversely effects the protoplasts which then have a reduced vitality. For this reason the enzyme treatment should be as short as possible.

Thought should also be given to the other isolation conditions: temperature, pH, light conditions, speed of shaking etc. Enzyme treatment is usually carried out in the dark.

After protoplasts are obtained they have to be purified. This is carried out by: filtration (through stainless steel or nylon sieves) to remove impurities such as cell walls and remains of the cells etc. (Fig. 24.4), followed by repeated centrifugation in a sugar solution to pellet the protoplasts. It is often impossible to pellet the protoplasts by centrifugation because they are too delicate, and they are then centrifuged at a low speed

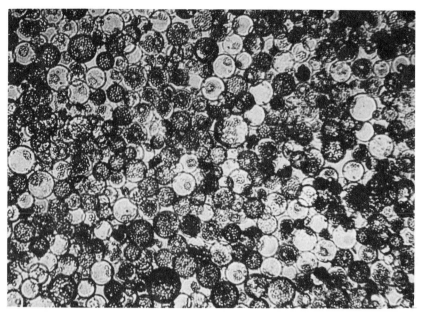

Fig. 24.4. Purified protoplasts of *Nicotiana tabacum.* Photograph supplied by Koornneef, Genetics Department, Agricultural University, Wageningen.

under which conditions the intact protoplasts float on a 15–20% sugar solution.

Fusion of protoplasts

To obtain efficient fusion it is necessary to work with a large number of viable protoplasts. Fusion is generally induced by incubation in a high concentration of PEG, a high concentration of Ca^{2+} and a relatively high pH (Menczel, 1983). This treatment is necessary to remove the negative charge on the protoplasts (Schilperoort, 1983). The PEG and Ca^{2+} ions are washed away after fusion has taken place. Electrofusion has become popular in the last few years (Zimmermann, 1982). Zimmermann has predicted great possibilities for this technique which results in more viable protoplasts than the use of PEG. Kameya (1984) realized the fusion of protoplasts with the use of dextran and electrical stimulus.

If different protoplasts fuse (originating from 2 different genotypes or plant species) then a heterokaryon results (Fig. 24.5) and if 2 similar protoplasts fuse then a homokaryon is obtained. After cell fusion has taken place nuclear fusion can occur when the nuclei divide simultaneously.

After somatic hybridization the biggest problem is to select the somatic hybrids from the cell population. If for example, hybridization is carried

266

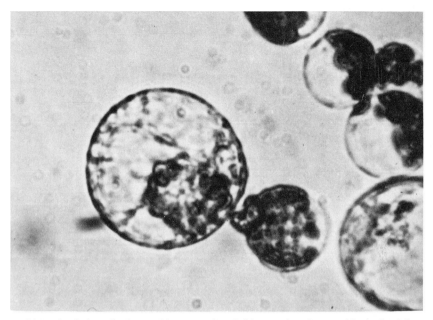

Fig. 24.5. The fusion of a large white protoplast (with cytoplasmic strands) obtained from a suspension culture of *Nicotiana tabacum*, with a small green (containing chlorophyll) protoplast from *Lycopersicon lycopersicum*. Photograph supplied by Koornneef, Genetics Department, Agricultural University, Wageningen.

out between protoplasts of two sorts A and B, then a population will be obtained which contains non-fused A and B protoplasts, AA and BB homokaryons, AB heterokaryons and diverse multiple karyons (fusions). Many selection procedures have been developed to efficiently select only the AB heterokaryons from this mixture and they will be dealt with in Section 24.7.

Many somatic hybrids have been produced in the last few years, especially among members of the *Solanaceae* (Pelletier and Chupeau, 1984).

The fact that somatic hybridization has been successful between *Solanum tuberosum* and *Lycopersicon lycopersicum,* between *Brassica* and *Arabidopsis* and between *Daucus carota* and *Aegopodium podagraria* shows that even intergeneric hybridizations are possible. This is especially interesting since these hybrids are impossible to obtain by sexual means (Vasil, 1978; Thomas et al., 1979). However the practical usefulness of these hybrids is extremely doubtful since they are sterile.

Somatic hybridization is interesting from an agricultural point of view, especially for important plants where it is possible to regenerate complete functional plants from protoplasts (Wenzel, 1983). Examples of these are

267

Brassica napus, Brassica campestris × *Brassica oleracea, Medicago sativa, Lycopersicon lycopersicum, Pennisetum americanum, Solanum tuberosum, Trifolium repens, Daucus carota, Nicotiana* species and *Petunia* species.

The culture of protoplasts

The culture of protoplasts has very specific requirements, because it consists of many different processes: regeneration of the cell wall, cell division, callus formation, and finally organ and/or embryo differentiation. During regeneration of the cell wall the Ca^{2+} concentration should be high and the (negative) osmotic potential of the nutrient medium low, whilst the cell density (as is also the case for cell suspension culture) plays an important role. After regeneration of the cell wall (within the first few days) the (negative) osmotic potential of the medium must be gradually increased (Burgess, 1983). The first cell divisions usually take place 2–7 days after protoplast isolation (Fig. 24.6).

The culture of protoplasts can take place in different ways (Vasil, 1980a; Ahuja, 1982). In solid media usually Agarose is used as the coagulating agent; SeaPlaque Agarose of Marine Colloids is strongly recommended as it is much less toxic than agar. Protoplasts are cultured as follows:

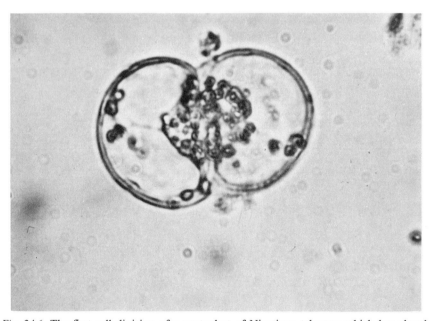

Fig. 24.6. The first cell division of a protoplast of Nicotiana tabacum which has already formed a cell wall. Photograph supplied by Koornneef, Genetics Department, Agricultural University, Wageningen.

1. Suspended in a thin layer of liquid medium in a Petri dish.
2. On solid media according to the Bergmann (1960) technique.
3. Grown in micro-chambers between glass plates, or in hanging drops.
4. In drops of liquid medium (micro-drop method), which are placed in a Petri dish.
5. In 'nurse cultures', especially when the density of the protoplasts is low. The feeder layer technique may be used instead of nurse culture; here the protoplasts are grown on a cell layer from which they are separated by cellophane or filter paper. Raveh et al. (1973) were the first to use the feeder layer technique using protoplasts of tobacco as the under layer, which had been irradiated with X-rays.
6. In a layer of (half) liquid medium which is on a solid medium.
7. Agarose-bead and agarose-disc culture (Shillito et al., 1983; Dons and Bouwer, 1986).

The culture of protoplasts with cell walls is very similar to that of individual cells (Chapter 20), requiring special attention to be paid to the composition of the nutrient medium and the physical growth factors (Binding et al., 1981). Protoplasts are grown mainly in the dark or at low irradiance. After the production of microcalluses (Fig. 24.7) subculture onto another medium is usual to induce regeneration.

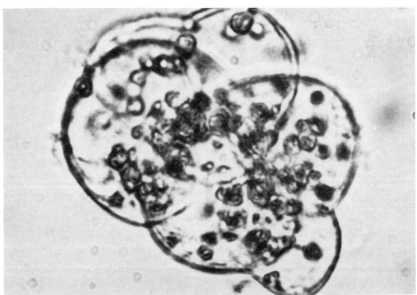

Fig. 24.7. Microcallus of *Nicotiana tabacum* consisting of about 8 cells, formed from a protoplast which has divided. Photograph supplied by Koornneef, Genetics Department, Agricultural University, Wageningen.

The regeneration of plants

The chance of regenerating a plant from isolated protoplasts is relatively small (Evans and Bravo, 1983; Binding et al., 1982). By 1983 it was only possible with 66 species of which the most species (38) belonged to the *Solanacea*: e.g. *Nicotiana, Petunia, Solanum, Lycopersicon, Atropa, Browallia, Capsicum, Datura, Hyocyamus*. The following species belong to other families: *Daucus carota, Manihot esculenta, Medicago sativa, Trifolium repens, Brassica napus, Asparagus officinalis* and *Citrus sinensis*. The *Gramineae* (except rice) are very difficult to regenerate (Potrykus et al., 1983).

The following factors determine the chances of success in regenerating a plant from protoplasts (Binding et al., 1982):

1. The family, the species and the genotype.
2. The state of differentiation of the donor cells; juvenile cells and cells from shoot tips have greater regenerative potential.
3. The growth conditions of the starting material.
4. The use of a relatively rich medium. The formation of callus from isolated protoplasts is often induced on Kao and Michayluk medium (1975), while the formation of embryos and organs can be carried out on B5-medium (Gamborg et al., 1968) to which 5–10% coconut milk is added. The Murashige and Skoog medium (1962) is also often used.

Details on the regulation of embryo and organ regeneration are given in Chapter 20.

24.5. The relevance of somatic hybridization

Below is a list of the possible advantages of somatic hybrids which has been compiled with the help of the following literature: de Groot et al., 1982; Thomas et al., 1979; Ingle, 1982; Schilperoort, 1984; Gleba and Sytnik, 1984; Hermsen, personal communications.

1. Using somatic hybridization it is possible to create hybrids which would not be possible by normal crossings, due to taxonomic or other barriers. An interesting example is the somatic hybridization between *Solanum tuberosum* (tuber forming, little or no disease resistance) and *Solanum brevidens* (non-tuber forming, disease resistant), which is not possible in vivo (Barsky et al., 1984). However, Hermsen already had many hybrids available with *S. brevidens* before Barsky had started his research and also had advanced material with *S. brevidens* char-

acteristics. This demonstrates that the classical plant breeder obtained hybrids in a round about way (a so-called bridging-cross), which were later also obtainable by somatic hybridization.

Two other examples of somatic hybridization, between *Brassica oleracea* and *B. campestris* (Schenck and Röbbelen, 1982) and *Medicago sativa* and *M. falcata* (Téoulé, 1983) suggested that amazing things could be achieved using this technique. However, these two somatic hybridizations are also possible by sexual means, although it is true that these crossings are difficult in vivo, and depend on the genotype.

It can be concluded that somatic hybridization is of no special advantage. If it is possible to obtain the hybrid by sexual means either directly or after a so-called bridging-cross, then in vitro somatic hybridization should be avoided, since somatic hybrids between incompatible species are usually unusable.

2. Somatic hybridization can be used as an alternative to obtain tetraploids, if this is unsuccessful with colchicine treatment (see also Section 25.2.3.).

   Making amphidiploids is also attractive to plant breeders, and somatic hybridization of two haploids can result in a diploid.

   In all these cases somatic hybridization is only valuable if the normal methods of mitotic doubling are impossible.

3. In principle, it is possible with somatic hybridization to make hybrids from plants which are sublethal, in which the sexual organs are abnormal or in which there is complete male sterility. However there is very little desire for these hybrids, and above all unwanted recessive characters are stored in the material which can re-appear after further crossings.

4. Hybridization becomes possible between plants still in the juvenile phase (not flowering) and those already in flower.

5. Partial hybridization can give rise to so-called asymmetric hybrids in which not all the chromosomes combine after nuclear fusion.

6. Usually during sexual reproduction only the nucleus (chromosomes) of the male gamete is (are) transferred to the progeny. With somatic hybridization both partners transfer nucleus and cytoplasm to the hybrid and progeny, which therefore also has hybrid cytoplasm.

   In this way it is possible to transfer genetic information to the progeny, which is localized in the cytoplasm of the male partner (e.g. male sterility, disease and herbicide resistance). With cytoplasmic hybridization the nucleus of one of the fusion partners should be eliminated, either before fusion (e.g. by irradiation) or after fusion (after somatic hybridization).

24.6. Disadvantages and problems of somatic hybridization

It is unrealistic to only consider the advantages and relevance of somatic hybridization without outlining some of the problems, disadvantages and technical difficulties that may be encountered:

1. The lack of an efficient selection method (often with the help of mutants; see Section 24.8.), is sometimes an insurmountable problem.
2. The end products after somatic hybridization are often unbalanced (sterile, missformed, unstable) and are therefore not viable, especially if the fusion partners are taxonomically far apart. This incompatibility can be caused by the choice of plant species (e.g. potato and

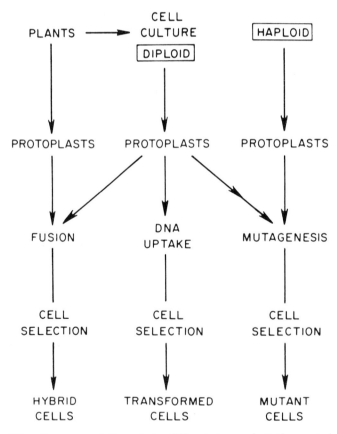

Fig. 24.8. Schematic representation of the different types of starting material (diploid, haploid) which are used for genetic engineering or manipulation (somatic hybridization, transformation, and mutation breeding). From Gamborg et al. (1974).

tomato) or the type of cell (e.g. mature cells from fruit tissue and juvenile tissue from a seedling).

3. The development of chimeric calluses, in the place of hybrids. This is usually due to the nuclei not fusing after cell fusion, and dividing separately. Plants which are regenerated from chimeras usually loose their chimeric characteristics, since adventitious shoots or embryos usually develop from a single cell.

4. Somatic hybridization of two diploids leads to the formation of an amphidiploid which is generally unfavourable (except when tetraploids are formed intentionally). For this reason, in most cases, the hybridization of 2 haploid protoplasts is normally recommended.

5. Regeneration products after somatic hybridization are often very variable, as a result of (Harms, 1983):
   — Chromosome elimination, which is especially common with hybridization of species which are taxonomically far apart so that the nuclei are incompatible. This results in e.g. aneuploids and even cybrids.
   — Translocation.
   — Somaclonal variation (See Chapter 21).
   — Organelle segregation. After the fusion of 2 cytoplasms so-called sorting of plastids takes place and also recombination of the mitochondria.

6. It is never certain that a particular characteristic will be expressed after somatic hybridization.

7. The genetic stability during protoplast, cell and callus culture is poor (See Chapter 21).

8. At the present time it is only possible with a small number of plant species to regenerate plants from protoplasts (see Section 24.4.). Protoplasts and cells of the *Gramineae* are especially difficult to regenerate except rice.

9. It is often assumed that somatic hybridization is possible in vitro for 2 species which are incompatible in vivo. It should be appreciated that plants are strongly inclined to retain their integrity by chromosome elimination (Cocking, 1977), both in vivo and in vitro.

10. The production of cybrids (cytoplasmic hybrids) has had little success. The best example where success has been achieved is the hybridization between *Nicotiana plumbaginifolia* (non-irradiated) with *Nicotiana tabacum* (irradiated utilizing radioactive cobalt). The nucleus of *N. tabacum* is eliminated, while the cytoplasm is still transferred to the new hybrid.

24.7. Selection procedures after somatic hybridization

There are many articles describing selection procedures (Widholm, 1978, 1982; Straub, 1976; Vasil, 1978; Schieder, 1980; Davey and Kumar, 1983; Ahuja, 1982; Pelletier and Chupeau, 1984), and a summary is given below. Gonzales and Widholm (1985) described wide ranging protocols for the selection of plants cells for desirable characteristics, especially inhibitor resistance. The plant breeder has the disadvantage that he is often selecting at the plant level rather than the cell level.

Selection is split into positive and negative selection. Negative selection is a method by which growing cells are preferentially killed and non-growing cells allowed to survive. With positive selection only growing cells are selected; most mutants obtained in plant cell cultures so far are the result of positive selection (Meridith, 1984).

The following are some of the most important selection procedures:

1. Direct isolation. When two protoplasts are seen under a microscope to fuse, then sometimes the resulting hybrid can be directly isolated and separated by using a micro-manipulator or micropipette. In practice this is difficult applicable since isolated protoplasts are very difficult to grow, and the 'nurse-method', the 'feeder-layer' or the 'micro-drop' methods need to be applied. Hein and Schieder (1984) described a new method which is based on the use of a micro-capillary with which the fusion products can be picked out.

2. Visual. Sometimes heterokaryons can be identified with the naked eye. With fusion of chlorophyll-less (with characteristic cytoplasmic strands) and chlorophyll containing protoplasts the hybrid cells are easily identified by the presence of cytoplasmic strands and chloroplasts (Fig. 24.5.).

   Another method of selection is to colour the 2 types of protoplasts with (different) fluorescent dyes. With the help of cell sorters the heterokaryons which contain both dyes can be selected. The cell sorter is a so-called continuous flow cytometer through which large numbers of fused and non-fused protoplasts pass a laser beam. The cells absorb light energy, a fraction of which is emitted as fluorescence. The fluorescence is measured and so the fusion products can be sorted out. Heterokaryons are then selected on the basis of their differential fluorescent colouring (Harkins and Galbraith, 1984).

3. Separation on the basis of physical characteristics. Physical methods are being developed by which hybrid cells can be selected. These can for instance, discriminate between the electric charge on the hybrid and that on both of the fusion partners (Vienken et al., 1981) or

between the effect of centrifugation on the hybrids in comparison to that on both the fusion partners. These methods however do not result in a complete selection of hybrids but result in a greater concentration of the hybrids.

4. Genetic complementation of recessive induced deficiencies. Here the hybrid AB will not possess the deficiencies shown by the fusion partners A and B. For example:

   — Hybridization between two light sensitive recessive mutants (SSvv) and sub-lethal (ssVV) of *Nicotiana tabacum* gives rise to a hybrid (SSssVVvv) which is not sensitive to light; surviving colonies are selected which form chlorophyll at high irradiance.

   — The hybrid between *Solanum nigrum* and *Petunia hybrida* can grow in a medium containing diethylpyrocarbamate (to which *S. nigrum* is sensitive) and iodine acetate (to which *P. hybrida* is sensitive).

   — Two cell lines of *Nicotiana sylvestris* are each resistant to a different chemical, the hybrid being resistant to both compounds.

   — *Nicotiana glauca* and *B. langsdorfii* both require auxin for growth whereas the somatic hybrid is an auxin autotroph.

   — Cell line A of the liverwort *Shaerocarpos donelli* requires nicotinic acid for growth and cell line B needs sugar since it is an albino. The hybrid formed grows without either nicotinic acid or sugar in the medium. The cell lines A and B are called auxotrophic mutant lines.

5. Heterosis effect. Hybrid cells which develop after the fusion between chlorophyll deficient protoplasts of *Datura innoxia* and green protoplasts of *Datura sanguinea* are characterized by better growth than cell colonies which develop from protoplasts of either fusion partner alone.

24.8. Selection of mutants

For the production and selection of somatic hybrids, cell-lines are often used with extreme characteristics (Ingram and Macdonald, 1986), which have arisen by spontaneous mutation (somaclonal variation, Chapter 21) or have been induced. Additionally mutants are used in plant breeding (disease resistance) and the biosynthesis of compounds in vitro (Fig. 24.8). The production of mutants is often far from simple and extremely time consuming. Generally a variant cell line is taken to mean that it is not known whether the mutation in this line has arisen spontaneously or as a result of epigenetic changes (Maliga, 1984).

Protoplast, cell and suspension cultures make possible screening of a large number of cells for desired mutations after mutation induction. Mutations can be induced chemically (ethylmethanesulphonate; N-ethyl-N-nitro-ureum), physically (X-rays or gamma irradiation) or by transformation with *Agrobacterium tumefaciens*. Haploid cells have the great advantage that recessive mutants are immediately visible (Maliga, 1984; King, 1984; Thomas et al., 1979; Straub, 1976; Vasil, 1978). Selection of mutants can take place in two ways:

1. Directly. The mutant cells have a selective advantage (positive selection) over the remainder of the cells. This is the most common type of selection and examples are resistance to salt, amino acids or their derivatives, bases of nucleic acids or their derivatives, antibiotics, water-stress, toxic substances, phytotoxins and heavy metals.
2. Indirectly. The non-mutated cells are selectively killed, because they are metabolically active under certain growth conditions, which are limiting for the growth of mutated cells. The selected mutants are then subcultured under enriched conditions which are not limiting for growth. This is known as the enrichment procedure.

   Auxotrophic cell-lines are an example of indirect selection of mutants. They grow very slowly on media which are deficient in an essential component and have normal growth where this is added to the medium.

In vitro methods for the selection of mutants offer several important advantages over their in vivo counterparts (Gunn and Day, 1986):

1. They can usually be applied at any time of the year.
2. They often make it possible to maintain closer control of the test conditions.
3. They avoid many hazards that are common to field and greenhouse tests.
4. They reduce the amount of space occupied by each test sample.

24.9. Transformation by Agrobacterium tumefaciens

The most natural way of genetic manipulation with the help of recombinant DNA can be carried out using the soil bacteria *A. tumefaciens* (the crown-gall bacteria). If dicotyledonous plants are infected by the crown-gall bacteria the plant cells are transformed and become hormone autotroph. The transformed cells have then acquired the characteristic of being able to produce auxin and cytokinin in such large amounts that tumours are formed. With the help of a few review articles (Wullems et al., 1982, 1986; Ooms et al., 1983; Dodds, 1985; Magnien and de Nettancourt, 1985; Gheysen et al., 1985; Hernalsteens et al., 1980) transformation by *A. tumefaciens* is described together with it's importance for gene transfer in higher plants.

The bacteria *A. tumefaciens* apart from the chromosomal DNA consists of a large plasmid which has been given the name Ti (Tumour inducing)-plasmid. If a plant cell becomes infected by the bacteria (it does not enter the cell), then a part of this plasmid (the so-called T-region) becomes build into the DNA of the plant cell by an as yet unknown mechanism. The bacterial DNA recombines with the chromosomal DNA in the plant nucleus. This foreign DNA becomes part of the plant genome and is called transferred DNA or T-DNA. There are a number of genes on the Ti-plasmid which are expressed within the plant: the onc or tumour genes. One gene is responsible for the over-production of cytokinin-like compounds, while two other genes are responsible for the production of auxin. If plant cells are transformed by *A. tumefaciens* then they become hormone autotroph as a result of over-production of cytokinin and auxin. If a mutation takes place at the auxin-locus of the Ti-plasmid then after transformation of the plant cells tumours with shoots are formed (so-called shooter-mutants). If the mutation is in the cytokinin-locus then tumours with a lot of roots are formed (rooter-mutants). Other genes on the Ti-plasmid are responsible for the tumour-specific opine (octopine and nopaline) synthesis of the transformed plant cells: opines are unusual amino acid derivatives which are produced by the crown-gall tissues.

As well as *A. tumefaciens* which induces tumours (crown-galls), A. rhizogenes causes hairy root disease on dicotyledons. *A. rhizogenes* consists of an Ri-plasmid part of which can also be integrated into the plant chromosome.

If DNA is to be carried by *A. tumefaciens* then it must first be built into the T-region of the Ti-plasmid. Wullems et al. (1986) state that a prerequisite for gene transfer is the isolation and characterization of genes of interest. The isolation of specific genes from fragmented eukaryotic chromosomes is still extremely difficult. Many complex traits that are of agricultural importance, such as yield and resistance to biotic and abiotic factors, are polygenic traits that are neither well identified nor biochemically characterized. Transformation of plants can be obtained by different methods (Wullems et al., 1982, 1986):

1. By the co-cultivation procedure (Marton et al., 1979). Bacteria are temporarely co-cultivated with protoplasts after they have regener-

ated a new cell wall; the transformation frequency with this procedure is very high (Gheysen et al., 1985). Unfortunately this procedure only works with dicotyledons, despite the first results being cited for monocotyledons (Magnien and de Nettancourt, 1985). When cell and/or protoplast populations are transformed by the crown-gall bacteria the problem arises of selecting these transformed cells. The selection usually makes use of the hormone autotrophic characteristic which results from the transformation by *A. tumefaciens*. Only hormone autotrophic cells are able to develop on a medium without hormones (or at the first selection on a medium with a low concentration of hormones). An interesting feature of co-cultivation is that it frequently produces octopine-like transformants with a high shooting

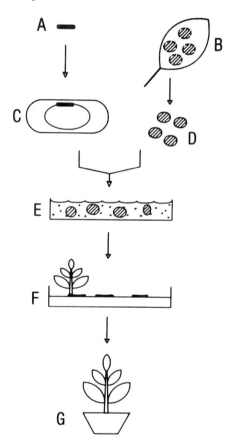

Fig. 24.9. Schematic representation of the transfer of an isolated gene into a plant by transformation with *Agrobacterium tumefaciens* (Dons and de Jong, 1986). A: new gene. B: leaf. C: new gene incorporated into the Ti-plasmid of *Agrobacterium tumefaciens*. D: leaf explants. E: leaf explants put into a suspension of bacteria. F: regeneration of a transformed shoot. G: plant with the new characteristic (gene).

capacity. In co-cultivation (holds also true for 2) experiments use is very often made of so-called disarmed Ti-plasmids from which the onc genes have been mutated or deleted; the advantage of this is that normal plants can be regenerated from the transformed cells.

2. An extremely important method is at the moment the so-called leaf-disc transformation-regeneration method (Horsch et al., 1985). To obtain infection the leaf discs are placed in a *A. tumefaciens* suspension (Fig. 24.9). If a transformed cell regenerates a new plant then the new gene(s) is (are) added to the plant. Using this relatively simple method there is a good chance that the adventitiously formed shoots will be transformed.

3. By application of the DNA transformation method (Draper et al., 1986) which is suitable for monocotyledons as well as dicotyledons. This method is designed particularly to avoid natural cellular interaction between *Agrobacterium* and dicotyledonous plants. The DNA can be offered in 4 different ways:
   — as pure DNA
   — as calcium phosphate-DNA co-precipitate
   — as DNA encapsulated in liposomes
   — by direct micro-injfection of DNA

4. Other bacteria and viruses can also be used as vectors for transfer of plant genes (Dodds, 1985).

5. Fusion of normal protoplasts and protoplasts from tumour tissues.

25. Miscellaneous applications

25.1. In phytopathology

25.1.1. *Outline*

It is now well known that in vitro culture of higher plants can be used in phytopathology (Ingram, 1973; Ingram and Helgeson, 1980; Miller, 1985; Wood, 1985). If protoplast, cell and callus cultures are used then it should be borne in mind that these cultures may differ in many respects (physiologically and sometimes genetically) from the intact plant.plant.

A list is given below of the areas in which in vitro culture of plants can be of help in phytopathology (Ingram and Helgeson, 1980; Miller and Maxwell, 1983; Miller, 1985; Wood, 1985):

1. One of the most important practical applications of in vitro culture in phytopathology is undoubtedly the production of plants free of viruses, fungi and bacteria (See Chapter 19).
2. In vitro culture of a parasite and host under controlled conditions with the elimination of other undesirable parasites has many advantages; host-parasite interactions can be studies efficiently in vitro. For example: the addition and uptake of labelled compounds can be readily achieved; only a limited number of host cells are needed for experiments; changes can easily be made in the host-parasite relationship by varying the nutrient medium; the number of host cells and the size of the inoculum can be controlled; inoculation is possible without wounding.
3. Utilization of in vitro culture provides a better insight into the behaviour of viruses in plant cells or tissues: how virus particles enter a protoplast or cell and become infectious; how viruses multiply inside protoplasts and cells; how the host reacts to virus infection; how the

virus particles move from cell to cell (Hildebrandt, 1977; Takebe, 1976; Wood, 1985).

The fact that it is much easier for virus particles to enter a protoplast than a cell, this was the immediate cause to use protoplasts as a model system to identify interactions between the virus and higher plants at a cellular level.

4. Nematology. In vitro culture is also being used more frequently in nematode research (Krusberg and Babineau, 1979).

5. Crown-gall. Tissue culturists have for a long time shown an interest in the in vitro growth of so-called 'crown-gall' tissues. In 1974 Braun gave detailed information in his handbook about *Agrobacterium tumefaciens,* which induces growth in plant tissues without the need for auxin and/or cytokinin. After infection of a dicotyledonous plant the cells become autonomous in their growth and start uncontrolled cell division; infection by *A. tumefaciens* can lead to the formation of roots (as is the case on addition of auxin) or shoots (as is the case on addition of cytokinin) without the addition of these regulators. In the last few years *A. tumefaciens* has played a very important role in genetic manipulation using recombinant-DNA; the Ti-plasmid of this bacterium can transfer genes which can then be expressed in another plant and transferred to the following progeny. Schilperoort (1984) has recently published an extensive review article which shows the enormous possibilities for genetic manipulation using *A. tumefaciens* (See Section 24.9).

Finally it should be realized that not only *A. tumefaciens* is effective for tumour induction (Thomas and Davey, 1975). This may be achieved as well by:

1. *Agrobacterium rhizogenes.*
2. Wound virus.
3. By crossing. If *Nicotiana glauca* is crossed with *N. langsdorffii* then so-called 'genetic tumours' develop, despite the fact that neither parent forms tumours.

6. Mutation induction can be used to obtain disease resistant plants. Some examples of this are given at the end of Section 25.2.4.

25.1.2. Transport of disease-free plant material

As disease-free plants are now more readily obtained by meristem culture followed by in vitro cloning, questions arose about the safe transport of this disease-free in vitro material (export, import and transport, under sterile conditions). Recently two review articles (Kahn, 1986; Parliman, 1986) have dealt with plant quarantine and the international shipment of tissue culture plants. They have discussed the following subjects:

1. Quarantine regulations and principles.
2. Natural and man-made pathways for the introduction of exotic plants.
3. Pest-risks analysis and safeguards.
4. Tissue culture as a safeguard.

The transport of disease free plant material is of importance both for the exporting and importing country. If the exporter guarantees that the in vitro grown material is disease-free and the importer verifies this, then few problems arise. It is vitally important that the soil is fully excluded as a source of undesirable infection, such as animals or micro-organisms. Verficiation that the in vitro culture is in fact free from all infections (including internal infections) is far from easy. Considering that the transport will probably take a number of days (e.g. from the Netherlands to Australia) then infections will almost certainly be detected by the quarantine authorities unless:

— Refrigeration during transport is available.
— Antibiotics are added to the medium.
— Activated charcoal is added to the medium.

If an exporter does not make use of these measures to counteract infection then the following can be concluded:

1. Only internally infected cultures will be imported without detection.
2. Test for internal infections are the only way of being certain that the material is disease-free (See Sections 9.3. and 9.4.).
3. A virus test is necessary to be absolutely certain that the plant material contains no known viruses (See Section 19.2.7.).
4. Despite all these precautions infections may still occur. One of the biggest causes of contamination is the rapid change in pressure experienced inside the container during air transport whenever a plane lands or takes off (Turner, 1983).

World-wide reactions have lead to the conclusion that disease-free transport is generally accepted as safe. However, the quarantine inspectors of Australia still require the fulfillment of many conditions, some of which are:

1. In vitro grown plant material also requires an import licence.
2. A number of plants (carnation, chrysanthemum, *Hydrangea, Pelargonium* and strawberry) may only be obtained from approved sources and must be accompanied by an endorsed disease-free certificate (Anonymous, 1980).
3. Plastic and glass containers as well as the agar must be clear with no turbidity since a visual inspection is then possible.
4. If saprophytes (which are not necessarily pathogenic) are present in the imported material, then this is for the inspectors an indication that the material is probably not disease-free.
5. Antibiotics should not be added to the medium in order to suppress the expression of micro-organisms present.
6. Some of the material transported may be required to stay in quarantine for up to 5 days.
7. The test tubes, plastic containers etc. should be firmly sealed; special conditions are required for closing test tubes, etc.
8. If living insects are found in a consignment (e.g. in the packing) then the whole consignment including the in vitro cultures is fumigated.
9. Samples may be taken over a period of time to establish the presence of antibiotics and micro-organisms.

Examples of plants transported disease-free are e.g. asparagus (Kahn, 1976) and *Citrus* (Button, 1977).

25.2. In plant breeding

25.2.1. *The development of chimeras in vitro*

There is little literature about the induction of chimeras by in vitro culture. Since in the past, especially when dealing with horticultural corps, chimeras were thought to arise due to grafting (the resulting plants were called graft-hybrids) it was anticipated that chimeras would be obtained with this technique in vitro. However, this is not the case.

Carlson and Chaleff (1974) first described the formation of chimeras in vitro. With adventitious shoot formation in callus of *Nicotiana tabacum* and *N. glauca x N. langsdorffii* 28 chimeras regenerated out of a total of 7,000 plants. Chlorophyll-chimeras were found to regenerate from callus of *Nicotiana tabacum* (Opatrny and Landa, 1974). Sree Ramulu et al. (1986) found that a high frequency of protoplast calli of potato are mosaic in composition, consisting of groups of variant (presumably mutated) cells: a majority of the mosaic calli gave rise to protoclones and mixtures of normal and variant protoclones; after tuber formation of the regenerated plants they found new plant types, which are probably due to chimerism, indicating the persistance of at least some chimerism.

In 1984 Marcotrigiano and Gouin (1984; 1984a), described the experimental formation of chimeras with tobacco, using a completely different technique to that of Carlson and Chaleff (1974). Marcotrigiano and Gouin (1984, 1984a) obtained chimeric callus by mixing cell suspensions which consisted of a mixture of cells (chlorophyllous and albino). These cell suspensions originated from green and albino cotyledons, hypocotyls, cells and cell suspensions respectively. When the chimeric calluses were induced to form adventitious shoots, of a total of 1,321 plants only 4 chimeras resulted. They explained the origin of chimeras as follows:

1. As a result of the experimental technique: mixing two types of cell suspensions, followed by the regeneration of adventitious shoots from more than one cell (Fig. 25.1.).

CALLUS TISSUE

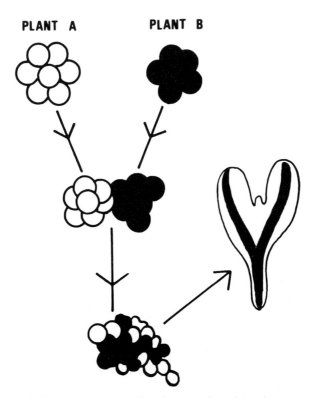

Fig. 25.1. Schematic illustration (hypothetic) of the way in which chimeras can be produced after mixing callus tissues of two plants.

2. As a result of spontaneous mutation, which often occurs in tissue culture. This explanation can certainly not be eliminated since also in the albino plants (the controls) which were formed from albino cells, green patches sometimes arose.

Finally, in other Chapters of this book it has been seen that chimeras can be obtained (e.g. by somatic hybridization), but in these cases the purpose of the experiments was not the production of chimeras. Transformation of plant tissues by *Agrobacterium tumefaciens* may also lead to chimeric plants.

It can be concluded that chimeras can be formed with the help of in vitro culture, although this has no significance in agriculture or horticulture. The question remains as to whether the formation of chimeras of ornamental plants in vitro is commercialy possible or desirable (Fig. 25.1).

25.2.2. *Separation of chimeras and isolation of mutants*

There are different explanations as to why a plant may have cells, sometimes even tissues, cell layers or cell sectors which are genotypically dissimilar.

1. Chimeras are undoubtedly one of the most important reasons. There are two types of chimera distinguishable, periclinal (in layers) and sectorial (in sectors).
2. During cell division, sometimes as a result of an error, mutated cells may arise, which may be retained and propagated or may be lost (See Chapter 21).
3. The development, either spontaneously or by mutation induction of mutated cells and from these the occasional development of the mutated sectors of plants.

It is often desirable that chimeras are separated into their constituent cell types or that mutants are isolated.

The separation of chimeras
Dommergues and Gillot (1973) and Johnson (1980) showed that the carnation cultivars White Sim and Jacqueline Sim (periclinal chimeras) could be separated in vitro. The principle behind this separation is adventitious shoot formation from a single cell layer. Existing chimeras were also separated with the help of in vitro culture with chrysanthemum (Bush et al., 1976), *Vitris vinifera* (Skene and Barlass, 1983), *Petunia*

(Pelletier and Delise, 1969; Bino et al., 1984) and *Euphorbia pulcherrima* (Preil and Engelhardt, 1982). Other examples of this are given in Chapter 21. Kameya (1975) isolated protoplasts from a chimeric *Pelargonium* and regenerated green as well as albino plants, showing that the growth of protoplasts can also be of use in separating chimeras.

Seeni and Gnanam (1981) showed using a chimeric heterozygous tomato (with green, greeny-yellow and white zones), that solid (non-chimeric) genotypes were regenerated by in vitro adventitious shoot formation. In this way it is possible by the use of adventitious shoot formation to loose the chimeric characteristic with tomato and to select a tomato genotype with increased photosynthetic activity. By traditional breeding methods it is not possible to effectively select tomato genotypes with an increased photosynthetic efficiency: this arises since the photosynthetic capacity is localized in the chloroplasts which are only maternally inherited. It should be borne in mind that the transfer of hereditable characteristics localized in the cytoplasm is, in principle, also possible by cybridization (Chapter 24).

Isolation of mutations
Sometimes mutations occur naturally, which if they are desirable (e.g. a new flower colour) can be isolated. Often the mutated part of the plant is so small that taking a cutting in vivo is impossible. This can be illustrated with *Kalanchoë blossfeldiana* (Karper and Pierik, 1981). In the inflorescence of different cultivars of this species mutated flowers sometimes arise spontaneously. If it is already known how to regenerate plants in vitro from flowers and flower stalks, then the isolation of a small mutated piece of tissue and the production of a mutant plant from it is possible. Karper and Pierik (1981) showed hw it was possible to obtain new cultivars using this technique (Fig. 25.2). In the same way, mutants can be isolated which are only present in a limited sector.

Another example of the isolation of mutants was given by Nickell (1973). Using sugar cane he showed that in callus and suspension cultures all kinds of deviating cells were present; by cell culture Nickell was able to isolate different cell lines and to regenerate plants with different numbers of chromosomes.

25.2.3. *Obtaining tetraploids and triploids; induction of chromosome loss*

1. Doubling the chromosome number (sometimes very desirable in plant breeding) is usually brought about by the addition of colchicine, an alkaloid isolated from the autumn crocus *Colchicum autumnale*.

Fig. 25.2. Part of the inflorescence of a salmonpink mutant of *Kalanchoë blossfeldiana* 'Roodkapje'. In the middle is one abnormal flower with petals. Three of these are yellow, one pink and one half pink/half yellow. Tissues isolated in vitro from the yellow section of the floret regenerate yellow plants (Pierik, Karper and Steegmans, unpublished results).

Seeds, shoot tips, or cuttings are dipped in a colchicine solution. The practice of this method in vivo often results in cytochimeras (Hermsen et al., 1981). The use of in vitro methods might reduce the frequency of chimeras drastically.

Plant species	Material	Conc. % (w/v)	Literature
Valeriana wallichii	Cell culture	0.05–0.20	Chavadej and Becker (1984)
Vaccinium species	Shoots	0.01–0.20	Lyrene (1984), Perry and Lyrene (1984), Goldy and Lyrene (1984)
Aranda 'Christine'	Shoot tip	0.05–0.075	Lee and Mowe (1983)
Phalaenopsis hybrids	Shoot tip	0.05	Griesbach (1981)
Saccharum officinale	Cell suspension	0.01	Liu and Shang (1974), Nickel (1973)
Dendrobium hybrids	Protocorms	0.01–0.05	Vajrabhaya and Chaichareon (1974)
Rubus species	Shoots	0.02–0.08	Gupton (1986)
Atractylodes lancea	Shoots	0.3	Hiraoka (1986)

Colchicine is first dissolved in water and since it is very thermolabile has to be filter-sterilized for in vitro use (Griesbach, 1981). The concentration used should not only be taken into account, but also the time taken for action: for prolonged treatment a low concentration of colchicine is required; whereas for short treatments higher concentrations are used (Perry and Lyrene, 1984). The concentration may vary from 0.01–0.3% (w/v); see table on page 288.

The induction of chromosome doubling is not always easy, as was seen by Wenzel and Foroughi-Wehr (1984) working with monoploid potatoes; however, the problem did not arise with diploid potatoes when colchicine was added.

2. Tetraploids are often formed after adventitious shoot formation, without treatment with colchicine. There are two possible explanations for the development of these tetraploids:
   — There are naturally tetraploid cells present in a diploid individual, from which complete (not chimeric) tetraploids can be obtained, because adventitious shoot formation usually occurs from a single cell (layer).
   — In adventitious shoot formation in vitro endomitosis (no separation of the chromosomes after cell division) can take place (possibly due to the use of cytokinin) resulting in tetraploids.

A few examples of plants where doubling of the chromosome number occurs after adventitious shoot formation are:

Nicotiana tabacum	Murashige and Nakano (1966)
Nicotiana tabacum 'Samson'	Dulieu (1972)
Brassica oleracea	Horak et al. (1975)
Daucus carota	Dudits et al. (1976)
Dendrobrium hybrids	Vajrabhaya (1977)
Kalanchoë 'Roodkapje' (Fig. 25.3)	Pierik and Karper (unpublished)
Lolium-Festuca hybrids	Rybczynsky et al. (1983)
Solanum tuberosum	Karp et al. (1984)
Solanum hybrids	Hermsen et al. (1981)

3. Polyploids can develop after somatic hybridization (Chapter 24). If for example the nuclei of two diploid protoplasts fuse, a tetraploid is formed.
4. Chromosome doubling in haploids. Kochar et al. (1971) and Nitsch et al. (1969a) showed that diploids occurred spontaneously in haploid tobacco. Kasperbauer and Collins (1972) isolated explants of the leaf middle nerve

Fig. 25.3. Left: a tetraploid plant of *Kalanchoë blossfeldiana* 'Roodkapje'. If adventitious shoots regenerate from leaf explants of this tetraploid, octaploids sometimes develop (right plant). From 130 regenerated plants grown, 8 were found to be octaploids (Pierik, Karper and Steegmans, unpublished results).

from haploid tobacco plants and diploids were formed after adventitious shoot formation. It was striking that this only occurred in the case of young leaves. It is also clear that fusion of nuclei at the first division of the pollen can lead to doubling of the chromosome number (Chapter 23).

5. Often a triploid can be obtained by crossing a diploid with a tetraploid. Triploids (usually sterile) can be very interesting to the plant breeder. In vitro triploids can be obtained by regenerating shoots from endosperm (Johri and Bhojwani, 1977). Full triploid plants of endosperm origin have been obtained with *Actinidia chinensis, Citrus grandis, Putranjiva rhoxburghii, Pyrus malus,* and *Santalum album* (Bhojwani, 1984).

6. Roth and Lark (1984) have recently shown that the addition of isopropyl-N(3-chlorophenyl)carbamate can result in the elimination of chromosomes, and that this substance dissolved in DMSO and added to cells of *Glycine max* resulted in the production of partially haploid cells.

290

25.2.4. *Mutation induction in vitro*

If seeds, plants parts with buds, and complete plants are irradiated (exposed to ionizing irradiations) to obtain mutants, chimeras are often formed. Chimeras are obtained because an irradiated meristem (with mutated as well as non-mutated cells) maintains it's intactness (Fig. 25.4) and a plant does not develop from only a single mutated cell. With a few exceptions chimeras are undesirable to the plant breeder. Mutation induction can be realized by chemicals (ethylmethanesulphonate; colchicine) or physical mutagens (irradiation: gamma, X-ray, etc.). The usual method is irradiation (Broertjes, 1981).

If complete mutants are wanted after mutation induction then the development of adventitious shoots is extremely important, since they usually originate from a single cell. Figure 25.4 shows what happens when a meristem is irradiated (a chimeric plant is formed), and a tissue is irradiated from which one mutated cell develops into an adventitious shoot (a solid mutant regenerates). However Norris et al. (1983) demonstrated that *Saintpaulia's* remain chimeric when cloned and therefore origin from more than one cell (layer); however, Broertjes and van Harten (1985) are in disagreement with the work of Norris et al. (1983).

Since adventitious shoot formation with many plants is impossible, or very difficult to induce in vivo, the plant breeder selecting for mutants has lately increasingly utilized irradiation and shoot regeneration in vitro. A tissue (leaf,

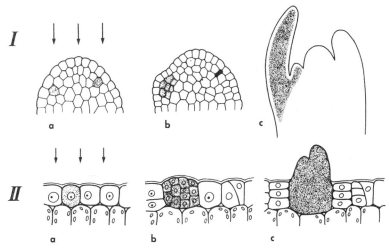

Fig. 25.4. Schematic illustration of the origin of a mutated shoot after irradiation. In I the shoot apex is irradiated, in II the epidermis (mutated cells are shaded). Ic shows a chimera (the left leaf and its dormant bud have been mutated). II shows the regeneration of an adventitious shoot (completely mutated) from one mutated epidermal cell. From Haccius and Hausner (1975).

petiole, peduncle, etc.) is usually irradiated before in vitro culture; the explants are then isolated in vitro and adventitious shoot formation induced (Fig. 25.5). After rooting of the shoots formed (Fig. 25.6), the plants are transferred into soil and then observed for mutations during their further development.

A good example of the above process is the chrysanthemum; Roest and Bokelman (1976) using irradiation treatment and in vitro culture techniques produced flower colour mutations. Roest (1980) and Broertjes and van Harten (1978) proposed that mutation induction in vitro can be a very useful tool for the plant breeder, since by irradiation and then adventitious shoot formation small (micro) mutations can often be obtained (size of the flower, flower colour, etc.) while other desirable characteristics are retained.

All the factors which influence mutation induction, especially in the case of vegetatively propagating plant species, cannot be dealt with in this Chapter, and these can be found in the handbook by Broertjes and van Harten (1978, 1987). This deals with such factors as dosimetry, sensitivity to irradiation, parts of the plant used, results to be expected for different plants etc.

During mutation induction only dominant mutants are usually directly visible. Recessive mutants can in most cases only be picked up after self

Fig. 25.5. The regeneration of adventitious shoots on a pedicel explant of chrysanthemum. Photograph from Roest and Bokelmann (1976). Photograph taken 8 weeks after isolation.

Fig. 25.6. The regeneration of adventitious roots at the base of a shoot cutting of chrysan-themum. Photograph from Roest and Bokelmann (1976). Photograph taken 4 weeks after isolation.

fertilization of corps propagated by seeds. However, in vegetatively propagated crops irradiation of heterozygous plants results directly in recessive mutants. The mutant Aa (obtained by irradiation of AA) gives AA, 2 Aa and aa after self fertilization, where aa but not Aa is seen as a mutant. If haploid cells are irradiated then the mutants are usually immediately discernable.

Little is known concerning the effect of irradiation on the regeneration potential in vitro. Bouriquet and Couvez (1967) showed that low doses of radiation did in fact promote regeneration (low dose stimulation), but higher doses strongly reduced regeneration. Pierik and Steegmans (1976b) showed with radiated callus of *Anthurium andreanum* in vitro that the regeneration potential strongly reduced with increasing radiation dose (Fig. 25.7). This was also found for irradiation of petioles in vitro. It appears that callus tissues are sometimes more resistant to irradiation than intact plants or cuttings. This was shown to be true for tobacco by Venketeswaran and Partanen (1966).

293

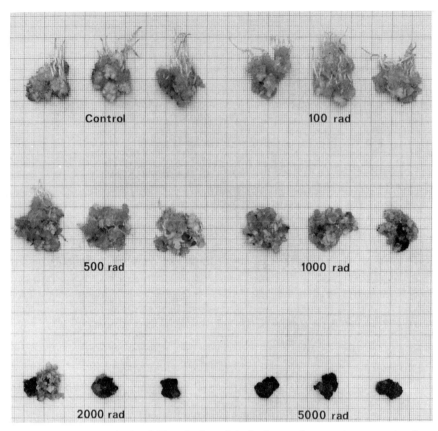

Fig. 25.7. The decrease in the ability of callus of *Anthurium andreanum* to regenerate adventitious shoots when the radiation dose is increased. At 100 rad the regeneration is slightly increased, but falls off at higher doses. At doses of 2,000 and 5,000 rad most calluses are killed (Pierik, 1976b). Photograph taken 10 weeks after isolation.

When mutations are identified in vitro after mutation induction and adventitious shoot formation, then it appears that groups of plants are produced which are mutated in the same way. This was found to be the case with mutation induction of *Kalanchoë blossfeldiana* (Pierik, Steegmans and Karper, unpublished data). This does not occur by chance, but is due to the hormone treatments applied. Taking *Kalanchoë* as an example: if cell A is mutated and under the influence of a high cytokinin concentration a completely mutated shoot (B) develops, then if the cytokinin concentration remains high the apical dominance is suppressed and axillary buds develop which are genetically identical to shoot B. Utilizing this in vitro system mutants are simply cloned.

It should be borne in mind that with mutation induction especially when callus is formed, but also sometimes when adventitious shoot formation is

294

induced, mutations can also arise spontaneously without the use of mutagens. This point is considered in relationship to somaclonal variation in Chapter 21. Mutants of one of our most important food crops, the potato, were obtained without mutagenesis (Shepard et al., 1980; Roest, 1980; van Harten et al., 1981).

Application
Here are a few examples where adventitious shoot formation in vitro is used in mutation breeding. Shigematsu and Matsubara (1972) and Matsubara (1974) obtained complete mutants of *Begonia rex* after irradiation and the formation of adventitious shoots in vitro. This was also shown for chrysanthemum and *Pelargonium* (Shigematsu, 1974). It was seen earlier in this Chapter that Roest and Bokelman (1976) carried out successful experiments with *Chrysanthemum* and Roest (1980) later described mutation induction in vitro for *Begonia x hiemalis*, chrysanthemum, carnation and potato. He showed that the optimal radiation dose (X-rays) varied from one species to another. For chrysanthemum, carnation, begonia, and potato these were: 8, 30–60, 15–20, and 15–27.5 Gy respectively. Van Harten et al. (1980) irradiated leaf explants of potato and obtained a high mutation frequency, a wide mutation spectrum and a very low rate of chimera production. A few recent examples of mutation induction are: *Saintpaulia, Streptocarpus* and *Kohleria* (Geier, 1983), *Saintpaulia* and *Pelargonium* (Grunewaldt, 1983) and *Euphorbia pulcherrima* (Preil et al., 1983).

It is obvious that other in vitro systems are used apart from adventitious shoot formation on explants, e.g. callus, suspension, cell suspension and protoplast cultures. A review of the theoretical aspects of mutation induction in cell and callus cultures can be found in Howland and Hart (1977) and Ingram and Macdonald (1986). The production of mutants in these systems are of importance in somatic hybridization (Chapter 24), the biosynthesis of compounds by cells and cell suspensions (Section 25.3) and the production of mutants that are of agricultural importance. Production of resistant strains is of special importance: resistance against diseases (especially toxins), salt, cold and heat, growth inhibitors, antibiotics, heavy metals and herbicides. Resistance at the cell level often differs completely to resistance at the plant level, or indeed may not be expressed at all at the plant level. Early flowering mutants are also extremely important, as are those with altered photosynthetic efficiency, yield, quality of the flowers etc., all of which are unfortunately extremely difficult to select at the cell level.

In agriculture and horticulture most important mutation induction research has been carried out on cell cultures and callus growth of *Citrus* and sugar cane (Broertjes and van Harten, 1978; 1987). In the case of sugar cane for example cell-lines and plants were obtained which had a higher sugar content (yield) (Heinz, 1973). Vardi et al. (1975) were the first to obtain mutations from irradiated protoplasts of *Citrus*. In 1975 they obtained whole plants from *Citrus* protoplasts. In Anonymous (1986: 93–192 and 391–449) special attention is paid to mutagenesis in vitro as a tool in mutation breeding. A great deal of mutation induction research was carried out with tobacco, where growth of protoplasts and regeneration of plants from calluses is relatively easy. Tobacco is especially interesting for mutation specialists because haploid cell-lines are available. Nitsch et al. (1969b) irradiated haploid tobacco plants and identified mutants directly. Schaefer (1974) obtained the same results with haploid tobacco when he used ethylmethanesulphonate as mutagen. Schieder (1976) also obtained mutants directly from irradiated haploid protoplasts of *Datura innoxia*.

Carlson (1973) first demonstrated that plant cells and protoplasts could be selected in culture for resistance to a pathogen toxin and that plants with an altered response to infection by the pathogen could be regenerated from these cultured cells. Daub (1986) in her review evaluated the many studies that used cell-culture selections to produce disease-resistant germplasm. A few striking examples of disease resistance which are obtained with the help of mutation induction in cell cultures are (Straub, 1976; Vasil, 1978; Wenzel, 1983; Yoshida and Ogawa, 1983; Daub, 1986):

Plant species	Resistance against
Brassica napus	*Phoma lingam*
Solanum tuberosum	*Fusarium oxysporum*
	Phytophthora infestans
Zea mays	*Cochliobolus heterostrophus*
	Helminthosporium maydis race T
Nicotiana tabacum	*Pseudomonas syringae pv. tabaci*
Saccharum officinale	*Helminthosporium sacchari*
	Drechslera sacchari

25.2.5. *Storage of plant material in vitro*

In recent years much interest has been shown in gene banks. It has been realized that extinction of different plant species (by excessive collectioning of protected species or by reclaiming and clearing of their natural environment, as well as the disappearance and dying out of old varieties) can be disastrous particularly for plant breeders. Generatively reproduced plants are simply kept for long periods in the form of seeds in closed containers or airtight packets. These are then (after the seeds are dried to a water content of 5–8%) stored at low temperature ($-18\,°C$ or lower) and low humidity. Only when problems occur with seed germination or when the seeds are only viable for short periods, other methods (in vitro) are considered for generatively propagated plants.

It is obvious that another solution is necessary for vegetatively propagated species. One such solution is a botanical garden, but this is not always economically feasible. If plants are kept under unprotected conditions, in the field, then diseases, pests and natural disasters can result in their loss. Since keeping them under strictly controlled conditions is not only expensive, but also in some cases impossible, other solutions have been found for the preservation of vegetatively propagated species. In the last 10 years in vitro culture (Fig. 25.8) has become very popular for this.

Fig. 25.8. Conservation of chrysanthemum in vitro at low temperature. Photograph taken by Engelhardt on the Bundesforschungsanstalt für gartenbauliche Pflanzenzüchtung in Ahrensburg, West Germany.

The in vitro system is extremely suitable for storage of plant material, since, in principle, it can be stored on a small scale, disease-free, and under conditions which limit growth. Much has already been written about gene banks and especially about storage at low temperatures. The following articles have been used in writing this Section: Withers and Street (1977), Bajaj (1979b, 1980b), Kartha (1981, 1985), Bajaj and Reinert (1977), Sakai (1984), Withers (1980, 1982, 1983, 1986, 1986a) and Henshaw (1982).

Withers (1982) recently published a review of the institutes world-wide which are concerned with in vitro storage of higher plants, and also reviewed the activities in this area taking place in different countries. She listed the different plants which were being studied and the problems which are encountered with storage in vitro (loss of regenerative potential, genetic instability). The storage of cells and organs in vitro at low temperatures was also described by Kartha (1985).

The requirements for in vitro storage are basically very similar to chose for vegetative propagation (Bajaj, 1980b; Withers, 1983; Henshaw, 1982):

1. Genetic stability should be maintained; this in generally the case with in vitro culture of meristems, shoot tips and embryos (Sakai, 1984).
2. Absence of disease must be guaranteed.
3. The regenerative potential should not be lost; this means that meristems, shoot tips and embryos are the safest material to use.
4. There must be little chance of damage or death of the material. Bajaj (1980b) proposed that organized structures such as meristems, shoot tips, and embryos were less likely to be damaged during e.g. freezing. However, Henshaw (1982) disagreed with this suggestion and claimed that homogeneous tissues, such as callus, are far more suitable for freezing than organized structures.

In conclusion, at this moment meristems etc. are preferred for the storage of plant material in vitro.

The conservation of plants in vitro has a number of advantages over that in vivo. These are as follows (Henshaw, 1974; Henshaw and O'Hara, 1984; Withers, 1980, 1983; Nitzsche, 1983; Bajaj, 1979a, 1980b):

1. In vitro culture enables plant species to be conserved that are in danger of being extinct.
2. In vitro storage of vegetatively propagated plants can result in great savings in storage space and time. Dale (1976) calculated that 800 cultivars of the grape vine could be stored in vitro in a few square meters, while this number in the field would occupy a hectare.

   It is often cheaper and easier to store meristems and shoots in vitro. In vitro storage of plant material is bound to become more important with an increase in fuel costs. The small scale conservation of parent-lines for the production of hybrid seeds is also extremely important for plant breeders.

3. Sterile plants which cannot be reproduced generatively can be maintained in vitro. Storage in vitro of material from a seed producing plant is also indicated when seeds are difficult to germinate, or loose their germination potential during normal storage.

4. It is possible in vitro to efficiently reduce growth, which decreases the number of subcultures necessitated.

5. If a sterile culture is obtained, often with great difficulty, then subculture in vitro is the only safe way of ensuring that it remains sterile. By maintaining the plant in vitro it is not necessary to attempt to produce plants from meristem cultures and then ensure that they are disease-free. Other problems are also often found when starting an in vitro culture (difficult to establish, necrosis, brown colouring, induction of rejuvenation) which might be encountered again if isolation has to be repeated.

6. Since continued sterile culture is possible then disease-free transport is also made possible (See Section 25.2.1.).

The following procedures are available to inhibit or stop development in vitro (Bajaj, 1979a, 1983a; Withers, 1980, 1983; Nitzsche, 1983):

1. Changing the composition of the nutrient medium thus limiting growth.

2. Dehydration by lowering the osmotic potential.

3. By lowering the atmospheric pressure, also known as hypobaric storage. The pressure of all the gases in the air is reduced. However it can also be accomplished by lowering the oxygen pressure alone; in this case the total pressure remains at 760 mm mercury but the oxygen is replaced by another gas e.g. nitrogen. Inhibition of growth by either of these procedures is applicable to both organized and nonorganized plant structures (Bridgen and Staby, 1983).

4. Sometimes inhibitors such as CCC and ABA are used. Mix (1984) used nodal explants of 220 potato cultivars as starting material for long term storage in vitro; plantlets were maintained at $10\,^{\circ}C$ in the presence of N-dimethyl-aminosuccinamid acid.

5. Storage at low temperature or by freezing. For short term storage a temperature of $2-5\,^{\circ}C$ is preferred for species from cold or temperate zones, and $8-15\,^{\circ}C$ for (sub)tropical species. If a longer storage period is necessary the material is usually frozen in liquid nitrogen $(-196\,^{\circ}C)$. The growth is severely inhibited at low temperatures and completely stopped upon freezing.

Freezing is far from easy since the formation of ice crystals can easily damage the material. If the freezing takes place very quickly then ice crystals can form intracellularly, whilst extracellular formation of ice crystals takes place if the freezing is carried out too slowly. Before freezing the cells, DMSO, glycerol, and sugars should be added to the nutrient medium (to remove water) otherwise they are unable to survive freezing.

Separate pretreatment and protocols have been described for the freezing and thawing of different plant material. More specific details can be found in the literature cited. It appears that freezing is, in principle, possible for all types of plant material: plantlets, embryos, shoot tips, meristems, callus, cell suspensions, pollen etc.

6. Combinations of the procedures given in 1-5 above.

Storage in vitro of meristem or shoot tip cultures has been successfully carried out with a large number of plants (Hu and Wang, 1983; Henshaw, 1982): carnation, strawberry, potato, pea, fruit trees, carrot, groundnut, *Prunus* species, *Ipomoea, Anthurium,* and *Gerbera.* Storage at low temperatures has certain advantages which have ensured that it is frequently used. Some of these advantages are:

1. Low temperatures slow the growth and development in a natural way, and limit the number of necessary subcultures. At a temperature of 2-5 °C subculture is only needed 1-2 times per year, and it is completely unnecessary when the material has been frozen.
2. The number of mutations is lower since these are more likely with each subculturing (Chapter 21).
3. It is easy to commercially plan and stagger production using low temperature techniques, making it possible, in principle, to have in vitro plant material available at any time of the year.
4. Haploid material can be conserved using low temperatures, whereas it quickly becomes diploid at higher temperatures.
5. Plant material that has been rejuvenated in vitro remains juvenile during cold storage, this being important for vegetative propagation.

Despite all these advantages in vitro storage is still not used on a large scale. It is by no means always as easy as is anticipated; it is also just as costly, and often gives rise to numerous technical problems. Each plant species, and in fact cultivars of a plant, all have their own specific requirements and their own problems. If callus, cell suspensions or similar material is stored then the genetic stability is often lessened, although this is not the case for storage of meristems, shoot tips or embryos.

25.3. The biosynthesis of substances in vitro

The culture of micro-organisms (*Penicillium, Streptomyces,* etc.) has long been used industrially to obtain substances such as penicillin, streptomycin, etc., which are important pharmaceutically. Higher plants are also an important source of all types of substances, especially medicins (glycosides, etherial oils, steroids etc.). Traditionally the medicinal plants have been grown and then the active components extracted, and this is likely to remain the normal procedure. However the production of medicinal plants can present problems, which have lead to the search for other ways to produce naturally occurring substances (Zenk, 1976; Alfermann and Reinhard, 1978):

1. Production in the field is strongly dependent on season, weather, climate, diseases and pests.
2. Naturally occurring sources, especially in the tropics and subtropical zones, are becoming limited and some medicinal plants are extremely scarce.
3. There may be technical and economic problems in production.
4. Production is labour intensive and therefore costs are high.
5. There may be political instability in the country where the plants are available resulting in an interrupted supply.

From considering the arguments put forward above, it can be seen that in analogy with micro-organisms, attempts have been made to obtain substances from cell suspension cultures of higher plants, either through accumulation in the cells (biomass) or sometimes by the release into the nutrient medium. If plant cells produce metabolites in vivo or in vitro which are not directly needed by the plant itself then these are termed secondary metabolites.

Much has been published on the production of secondary metabolites in vitro. The following articles have been consulted when writing this Chapter: Zenk (1976), Reinhard (1974), Alfermann and Reinhard (1978), Schilperoort (1983), de Groot et al. (1982), Mantell and Smith (1984), Barz et al. (1977), Douglas et al. (1979), Butcher (1977), Staba (1977, 1980, 1985), Reinhard and Alfermann (1980), Stohs (1980), Böhm (1980), Morris et al. (1985), Neumann et al. (1985), Fowler (1986).

Two main approaches have been followed in connection with the production of secondary metabolites (Morris et al., 1985):

1. The rapid growth of suspension cultures in large volumes which are subsequently manipulated to produce secondary metabolites.
2. The growth and subsequent immobilization of cells which are used for the production of compounds over a prolonged period.

For the vitro biosynthesis of substances plants are, in principle, selected which have a high production of the desired metabolite. Consequently organs or tissues are used in which the production of the metabolite is localized or is present at the highest concentration. After callus induction on explants, a cell suspension culture is set up; this culture has to be regularly selected for high-producing cell-lines. It must be taken into account that mutations occur regularly in cell cultures which can result in a reduced yield of the metabolite. Mutation induction has been used in an attempt to obtain high-producing cell lines. The production of secondary metabolites is strongly dependent on the rate of cell division; it is therefore not surprising that the yield of metabolites is dependent on a large number of factors (Mantell and Smith, 1984; Tabata, 1977; Dougall, 1979):

1. The starting material.
2. The pre-treatment before in vitro culture.
3. Physical growth factors: light, temperature, aeration, etc.
4. The composition of the medium.
5. Sometimes a so-called 'two-stage culture system' is necessary. The first stage involves growing the cells on maintenance medium and the second stage involves transferring the cells to a production (of secondary metabolites) medium (Morris et al., 1985).

The development of biosynthetic production in vitro has been extremely rapid because:

1. There has been a gradual introduction of growth and production from plant cells in fermentors and bioreactors (Fowler, 1984).
2. The growth of plant cells can be optimized by changes in the nutrient medium and the physical growth factors (aeration, stirring); also it has been attempted to increase biosynthesis by the addition of precursors.
3. Immobilized plant cells can be used (packed in a jelly-like mass) to increase production of the products and also to accumulate the metabolites in the medium. This accumulation means that substances excreted by the cells can be obtained by simply exchanging the medium. Recent research is covered by: Lindsey and Yeoman (1985, 1986), Rosevear and Lambe (1985), Berlin (1985) and Morris et al. (1985).
4. The interest in the in vitro production of other compounds (e.g. biodegradable nematicides and insecticides) increases. It is hoped that it will be possible to produce substances in vitro which are impossible to normally biosynthesize with plants in the field. With this in mind it is hoped that genetic manipulation of cells can result in them gaining the characteristic required of being able to produce a particular substance.

5. Selection and screening techniques have been developed for the growth of plant cells which is hoped will result in higher production of secondary metabolites (Berlin and Sasse, 1985).

6. Biotransformation is becoming more important. This is a technique which utilizes enzymes located in the plant cells to alter the functional group chemistry of externally supplied chemical compounds (Morris et al., 1985). There are two types of biotransformation:
   — via whole cells (immobilized or non-immobilized).
   — via the use of immobilized compound preparations.
   A good example of biotransformation is the conversion of digitoxin to digoxin by cells of *Digitalis lanata.*

7. Interest has grown in the increase in accumulation of secondary metabolites by the use of elicitors (Constabel and Eilert, 1986). Elicitors are strictly speaking compounds of biological origin involved in plant-microbe interaction. Elicitors such as phytoalexins (biotic elicitors) which are mediator compounds of microbial stress or stress agents such as osmotic pressure or heavy metal ions (abiotic elicitors) are used to increase accumulation of products in plant cel cultures.

8. Multiple shoot cultures are becoming a viable alternative in in vitro systems for the production of plant constituents (Heble, 1985).

Undoubtedly the biosynthesis of compounds in vitro has many possibilities, although a number of the problems associated with it still need extensive research before this new technique can be applied on a large scale. Some of these problems are:

1. Plant cells have a very long doubling time (16–24 h) when compared to micro-organisms (approximately 1 h). This means that plant cells produce relatively little biomass and therefore only small amounts of secondary metabolites.

2. If the production of se secondary metabolites in the intact plant is restricted to a particular organ, for instance to a differentiated tissue (with e.g. laticifers or glands), then little production takes place in the wells in vitro; the cells in vitro are often non-differentiated.

3. The production of secondary metabolites in vitro is generally very low, certainly when compared to an intact plant.

4. Cell and suspension cultures are genetically unstable; through mutation the production of compounds often falls to a level lower than that at the start of the in vitro culture.

5. Plant cells have the peculiar characteristic of forming aggregates in vitro (to clump). Vigorous stirring is needed to prevent this clumping, but this may result in damaging the cells. These problems can be avoided by the

use of the so-called 'air-lift fermentors' (passage of air bubbles keeps the cells in motion and fragments aggregates).

6. The use of a high sugar concentration in the nutrient medium (up to 5% w/v) increases the costs considerably.
7. Infections can considerably limit the process. When a bacterial infection occurs it quickly spreads over the culture as a result of it's shorter doubling time.
7. Much more intensive research is needed into bioreactors and the development of good methods of isolation.
8. Lack of knowledge about the growth of plant cells in bioreactors and fermentors is severely hindering their further development.

That there are still a great number of problems associated with the production of secondary metabolites is clearly illustrated by the fact that there is only one clear chemical success to-date (Staba, 1985). This sucless has been utilized by Mitsui Petrochemical Industries (Tokyo, Japan) who have commercially produced shikonin (a dye and antimicrobial compound, valued at approximately US $ 4,000 per kg), from *Lithospermum erythorhizon* plant cells in 750-litre fermentors.

26. In vitro cloning of plants in the Netherlands

Total number of plants cloned in vitro in the Netherlands in:

1985	35,981,960
1986	42,753,600

Increase from 1985–1986 is 19%

List of plants propagated in the Netherlands in numbers greater than 100,000 in 1986:

Bulbous and cormous crops	:	*Lilium*
Orchids	:	*Cymbidium*
		Vuylstekeara
Cut flowers	:	*Gerbera*
		Anthurium andreanum
Pot plants	:	*Nephrolepis*
		Saintpaulia
		Anthurium scherzerianum
		Cordyline
		Davallia
		Synchonium
		Bromeliaceae
		Alocasia
		Ficus
		Spathiphyllum
Plants for aquaria	:	*Cryptocoryne*
		Echinodorus
Agricultural crops	:	*Solanum tuberosum*

Pot plants	1982	1983	1984	1985	1986
Nephrolepis	5,108,750	7,921,630	8,306,700	10,101,350	11,194,900
Saintpaulia	4,975,250	5,371,250	4,400,000	4,389,904	3,715,000
Anthurium scherz.	37,500	269,040	400,000	898,700	1,747,000
Cordyline	42,100	493,300	578,150	604,500	783,000
Synchonium	51,950	9,300	162,550	242,490	603,500
Spathiphyllum	45,000	13,200	10,300	35,450	512,700
Davallia	95,000	258,000	643,000	570,700	340,000
Bromeliaceae	97,700	278,350	224,080	163,925	250,500
Alocasia	—	—	3,200	—	180,000
Ficus	314,500	390,000	330,000	161,500	148,000
Leea	—	—	—	—	55,000
Hortensia	47,600	124,000	145,000	105,000	51,000
Gardenia	—	—	20,000	25,000	40,000
Cryptanthus	—	—	—	—	38,504
Platycerium	—	—	—	—	36,200
Zantedeschia	—	—	—	—	24,000
Philodendron	4,520	6,341	180	29,550	16,500
Caladium	17,300	20,000	—	—	16,000
Dieffenbachia	—	—	850	750	14,000
Bouvardia	250	—	2,000	2,500	10,000
Calathea	—	—	—	240	10,000
Alternanthera	—	—	—	—	7,000
Anigozanthos	—	—	—	—	5,600
Begonia	64,000	19,000	7,500	10,747	5,000
Pellaea	—	—	—	—	4,500
Aralia	2,420	16,750	36,515	55,315	4,480
Pelargonium	200	—	—	100	3,330
Smithiantha	—	—	130,400	9,800	3,000
Nepenthes	—	—	575	—	1,300
Kalanchoë	250	500	3,500	700	1,200
Hypoestes	—	—	—	—	1,000
Yucca	170	11,666	950	2,680	60
Aphelandra	—	—	—	1,000	—
Dracaena	—	1,000	1,500	500	—
Pandanus	—	—	30	185	—
Cactaceae	16,700	32,000	15,000	—	—
Primula	5,000	—	4,000	—	
Monstera	—	—	2,140	—	—
Schefflera	1,200	—	10	—	—
Streptocarpus	4,700	5,000	—	—	—
Achimenes	275,000	3,000	—	—	—
Aloë	1,000	—	—	—	—
Maranta	—	—	—	—	—
Croton	—	—	—	—	—
Total	11,208,260	15,243,327	15,428,130	17,412,586	19,822,274

Orchids	1982	1983	1984	1985	1986
Cymbidium	1,051,100	667,700	507,000	793,500	1,053,790
Other orchids	862,100	679,450	1,027,500	323,240	395,400
Total	1,913,200	1,347,350	1,534,500	1,116,740	1,449,190

Cut flowers	1982	1983	1984	1985	1986
Gerbera	2,163,046	3,510,579	9,698,854	11,128,202	12,068,614
Anthurium andreanum	96,570	78,568	273,106	165,997	326,460
Alstroemeria	11,000	27,000	33,000	55,500	79,220
Chrysanthemum	3,119	11,350	27,450	67,125	73,650
Rosa	–	–	–	–	50,000
Statice	–	–	1,500	2,000	40,314
Carnation	1,500	7,000	2,500	2,000	1,500
Matricaria	–	–	580	–	–
Total	2,275,235	3,634,497	10,036,990	11,420,824	12,639,758

Vegetables	1982	1983	1984	1985	1986
Cauliflower	22,980	24,635	30,212	36,705	47,666
White cabbage	–	–	12,500	15,500	6,200
Red cabbage	–	–	12,500	12,500	6,000
Broccoli	500	–	600	1,000	5,104
Leek	–	620	2,100	3,900	2,000
Brussels sprouts	150	900	50	–	1,000
Tomato	–	–	100	300	500
Lettuce	–	–	100	485	200
Kohlrabi	–	103	60	150	100
Fennel	–	–	–	–	55
Red sugar loaf	–	–	–	355	–
Cucumber	–	–	–	310	–
Chicory	–	–	1,243	–	–
Beetroot	–	200	200	–	–
Asparagus	–	–	150	–	–
Sweet pepper	–	–	100	–	–
Aubergine	–	–	100	–	–
Carrot	–	–	25	–	–
Cichory	1,740	495	–	–	–
Onion	100	100	–	–	–
Total	25.470	27,053	60,040	71,205	68,825

Ornamental bulbs and corms	1982	1983	1984	1985	1986
Lilium	802,585	579,914	1,418,472	5,303,549	8,020,133
Nerine	10,680	3,105	5,910	3,363	40,500
Hyacinthus	1,000	3,007	9,584	20,828	13,234
Narcissus	—	—	500	21,000	10,000
Hippeastrum	1,500	—	1,000	2,000	1,150
Ornithogalum	40,750	—	280	—	506
Freesia	5,295	2,366	11,213	8,000	210
Gladiolus	11,000	2,000	200	—	155
Watsonia	—	—	—	—	32
Crindonna	—	—	10,000	—	—
Scilla	—	—	750	—	—
Vallota	—	—	110	—	—
Total	872,810	590,392	1,458,019	5,358,740	8,085,920

Miscellaneous	1982	1983	1984	1985	1986
Cryptocoryne	—	—	30,000	150,000	154,000
Echinodorus	—	—	2,000	62,000	148,000
Gentiana	—	972	7,933	13,618	26,500
Anubia	—	—	—	—	11,000
Hosta	200	—	9,120	5,365	9,700
Syringa	—	—	—	2,750	8,000
Rhododendron	—	—	3,000	8,000	5,400
Campanula	—	850	130	—	5,140
Strawberry	—	—	—	568	3,000
Kalmia	—	—	80,000	2,000	1,600
Bergenia	—	—	—	14	1,425
Stokesia	—	—	—	—	1,000
Betula	—	—	—	850	740
Populus	—	—	90,000	50,000	300
Thalictrum	1,200	7,130	20,200	4,300	—
Photinia	—	—	—	1,200	—
Nandina	—	—	—	800	—
Aristolochia	—	—	—	400	—
Sequoia sempervirens	2,000	—	—	—	—
Actinidia chinensis	—	—	—	—	—
Total	3,400	8,952	242,383	301,865	375,805

Agricultural crops	1982	1983	1984	1985	1986
Potato	156,847	256,900	260,000	280,000	286,825
Sugarbeet	9,500	22,500	20,000	20,000	25,000
Total	166,347	279,400	280,000	300,000	311,828

27. Index

318

28. Literature cited

In the literature list there are some unusual journals (in the Dutch language), especially for foreigners. Therefore, the following explanations are necessary:

Bloembollencultuur = a Dutch weekly journal for bulb growers without English summary.

Gewasbescherming (Gewasbesch.) = a Dutch journal for phytopathologists without English summary.

Groenten en Fruit = a Dutch weekly journal for vegetable specialists without English summary.

Jaarverslag Instituut voor Tuinbouwgewassen (Jaarverslag IVT) = Annual report of the Institute for Horticultural Plant Breeding, Wageningen, the Netherlands, without English summary.

Landbouwkundig Tijdschrift (Landbouwk. Tijdschr.) = a Dutch agricultural journal with English summary.

Mededelingen Directie Tuinbouw (Meded. Dir. Tuinbouw) = a Dutch journal for horticulturists without English summary.

Natuur en Techniek = a Dutch monthly journal on nature and technique, without English symmary.

Orchideeën = a Dutch journal for orchid specialists without English summary.

Vakblad voor Biologen = a Dutch journal for biologists without English summary.

Vakblad voor de Bloemisterij (Vakbl. Bloem.) = a Dutch weekly journal for floriculturists without English summary.

Verslagen Landbouwkundige Onderzoekingen (Agricultural Research Reports) = a Dutch series of agricultural publications (most of them in English), published by Pudoc, Wageningen, the Netherlands

Zaadbelangen = a Dutch journal for seed producers without English summary.

Aartrijk, van. 1984. Adventitious bud formation from bulb-scale explants of *Lilium speciosum* Thunb. in vitro. Dissertation, Agricultural University, Wageningen, the Netherlands: 1–79
Aartrijk, van et al. 1986. In: Withers and Alderson. 1986: 55–61
Abo-El-Nil and Hildebrandt. 1971. Plant Dis. Reptr. 55: 1017–1020
Adams. 1972. J. Hort. Sci. 47: 263–264
Aerts. 1979. Meded. Fac. Landbouwwet. Gent 44: 981–991
Aghion-Prat. 1965. Physiol. Vég. 3: 229–303
Ahuja. 1982. Silvae Genet. 31: 66–77
Albersheim and Darvill. 1985. Scient. Am. 253(3): 44–50
Albersheim et al. 1986. HortSci. 21: 842
Alfermann and Reinhard. 1978. Production of natural compounds by cell culture methods. Gesellsch. Strahlen und Umweltforschung, München: 1–361
Alfermann et al. 1980. Planta Medica 40: 218
Allen. 1974. Acta Hort. 36: 235–239
Allenberg. 1976. Die Orchidee 27: 28–31
Ammirato. 1983. In: Evans et al. 1983: 82–123
Ammirato. 1986. In: Withers and Alderson. 1986: 23–45
Ammirato. 1986a. HortSci. 21: 843
Ammirato et al. 1984. Handbook of plant cell culture. 3. Crop species. MacMillan Publ. Comp., New York: 1–620
Ando et al. 1986. Scientia Hort. 29: 191–197
Andreassen and Ellison. 1967. Proc. Amer. Soc. Hort. Sci. 90: 158–162
Anonymous. 1978. Plant tissue culture. Pitman, Boston: 1–531
Anonymous. 1978a. In vitro multiplication of woody species. Round Table Conf., Gembloux, Belgium, Centre Rech. Agron. Etat: 1–295
Anonymous. 1980. Seed Nursery Trader 78: 7
Anonymous. 1980a. C.R. Séanc. Acad. Agric. France 66: 619–719
Anonymous. 1984. Michigan Biotech. Inst. 1984: 1–194
Anonymous. 1986. Nuclear techniques and in vitro culture for plant improvement. Intern. Atomic Energy Agency, Vienna: 1–529
Appelgren and Heide. 1972. Physiol. Plant. 27: 417–423
Arditti. 1967. Bot. Rev. 33: 1–97
Arditti. 1977. Orchid Biology. Reviews and perspectives I. Cornell Univ. Press, Ithaca: 1–310
Arditti. 1977a. Orchid Review 85: 102–103
Arditti. 1982. Orchid Biology. Reviews and perspectives II. Cornell Univ. Press, Ithaca: 1–390
Arditti and Ernst. 1982. In: Stewart and Merwe, van der. 1982: 263-277
Arditti and Harrison. 1977. In: Arditti. 1977: 157–175
Arditti et al. 1982. In: Arditti. 1982: 243-370
Asjes et al. 1974. Acta Hort. 36: 223–228
Assche, van. 1983. Technische en economische aspecten van de weefselteelt bij planten. Eindwerk, Kath. Univ. Leuven, Belgium: 1–134
Augé et al. 1982. La culture in vitro et ses applications horticoles. Lavoisier, Paris: 1–152

Bagga et al. 1986. Int. Congr. Plant Tissue Culture Abstr. 6: 276
Bajaj. 1974. Euphytica 23: 633–649
Bajaj. 1979a. Euphytica 28: 267–285
Bajaj. 1979b. In: Sharp et al. 1979: 745–774
Bajaj. 1980. In: Rao et al. 1980: 50–67

Bajaj. 1983. In: Evans et al. 1983: 228-287

Bajaj. 1983a. In: Sen and Giles. 1983: 19-41

Bajaj. 1986. Biotechnology in agriculture and forestry 1. Trees I. Springer Verlag, Berlin: 1-515

Bajaj. 1986a. Biotechnology in agriculture and forestry 2. Crops I. Springer Verlag, Berlin: 1-608

Bajaj and Bopp. 1971. Ang. Bot. 45: 115-151

Bajaj and Gosal. 1981. In: Rao. 1981: 25-41

Bajaj and Pierik. 1974. Neth. J. Agric. Sci. 22: 153-159

Bajaj and Reinert. 1977. In: Reinert and Bajaj. 1977: 757-777

Baker and Phillips. 1962. Phytopath. 52: 1242-1244

Balatkova et al. 1977. Plant Sci. Lett. 8: 17-21

Ball. 1946. Am. J. Bot. 33: 301-318

Ball. 1950. Growth 14: 295-325

Ball. 1953. Bull. Torrey Bot. Club 80: 409-411

Bannier and Steponkus. 1972. Hortscience 7: 194

Barclay. 1975. Nature 265: 410-411

Barker. 1969. Can. J. Bot. 47: 1334-1336

Barlass et al. 1982. Ann. Appl. Biol. 101: 291-295

Barsky et al. 1984. Plant Cell Rep. 3: 165-167

Barz et al. 1977. Plant tissue culture and its biotechnological application. Springer Verlag, Berlin: 1-419

Bastiaens. 1983. Meded. Fac. Landbouwwet. Gent 48: 1-11

Bastiaens et al. 1983. Meded. Fac. Landbouwwet. Gent 48: 13-24

Bayliss. 1980. In: Vasil. 1980: 113-144

Beasly et al. 1974. In: Street. 1974: 169-192

Beauchesne. 1982. In: Earle and Demarly. 1982. 268-272

Beauchesne et al. 1970. Physiol. Plant. 23: 1101-1109

Benbadis. 1968. Coll. C.N.R.S. Paris: 121-129

Bergman. 1972. Am. Hort. 51: 41-44

Bergmann. 1960. J. Gen. Physiol. 43: 841-851

Bergmann. 1967. Planta 74: 243-249

Berlin. 1985. Newsletter I.A.P.T.C. 46: 8-14

Berlin and Sasse. 1985. Adv. Biochem. Engin. Biotechn. 31: 99-132

Bernard. 1909. Ann. Sci. Bot. 9: 1-196

Bhojwani. 1984. In: Vasil. 1984: 258-268

Bhojwani and Razdan. 1983. Plant tissue culture: theory and practice. Elsevier, Amsterdam: 1-502

Bhojwani et al. 1986. Plant tissue culture, a classified bibliography. Elsevier, Amsterdam: 1-789

Bigot and Nitsch. 1968. C.R. Acad. Sci. 267: 619-621

Binding. 1974. Z. Pflanzenphysiol. 74: 327-356

Binding et al. 1981. Z. Pflanzenphysiol. 101: 119-130

Binding et al. 1982. In: Fujiwara. 1982: 575-578

Bino et al. 1984. Heredity 52: 437-441

Bitters et al. 1972. Citrograph 57: 85-86 and 105

Blake. 1983. In: Dodds. 1983: 29-50

Böhm. 1980. Intern. Rev. Cytol. Suppl. 11B: 183-208

Bomhoff et al. 1976. Molec. Gen. Genet. 145: 177-181

Bonga. 1982. In: Bonga and Durzan. 1982: 387-412

*Bonga and Durzan. 1982. Tissue culture in forestry. M. Nijhoff and W. Junk Publ., The Hague, the Netherlands: 1–420

Bottino. 1984. In: Vasil. 1984: 13–17

Boulay. 1986. Int. Congr. Plant Tissue Cell Culture Abstr. 6: 9

Bouniols. 1974. Plant Sci. Lett. 2: 363: 371

Bouniols and Margara. 1968. Ann. Physiol. Vég. 10: 69–81

Bourgin and Nitsch. 1967. Ann. Physiol. Vég. 9: 377–382

Bouriquet. 1952. C.R. Soc. Biol. 146: 1897–1899

Bouriquet. 1972. C.R. Acad. Sci. 275: 33–34

Bouriquet and Couvez. 1967. Bull. Soc. Bot. France 114: 61–72

Bouriquet and Vasseur. 1965. Les phytohormones et l'organogenèse. Int. Coll. Liège: 381–389

Bourne. 1977. In: Fossard, de. 1977: 81–90

Boxus. 1973. Acta Hort. 30: 187–191

Boxus. 1974. J. Hort. Sci. 49: 209–210

Boxus et al. 1984. In: Ammirato et al. 1984: 453–486

Braak. 1967. Jaarverslag IVT Wageningen, the Netherlands: 14

Braak. 1968. Jaarverslag IVT Wageningen, the Netherlands: 14–15

Bradley et al. 1985. In: Henke. 1985: 307–308

Bragt, van et al. 1971. Misc. Papers Landbouwhogeschool Wageningen, the Netherlands 9: 1–147

Brainerd and Fuchigami. 1984. J. Environm. Hortic. 1: 23–25

Braun. 1974. The biology of cancer. Addison Wesley, Reading: 1–169

Bridgen and Staby. 1983. In: Evans et al. 1983: 816–827

Bright et al. 1985. Cereal tissue and cell culture. Martinus Nijhoff/W. Junk Publ., Dordrecht, the Netherlands: 1–304

Brocke, von. 1972. Diss. Everhard-Karls-Universität, Tübingen: 1–234

Broertjes. 1981. Natuur en Techniek 49: 420–439

Broertjes and Harten, van. 1978. Application of mutation breeding methods in the improvement of vegetatively propagated crops. Elsevier, Amsterdam: 1–316

Broertjes and Harten, van. 1987. Applied mutation breeding for vegetatively propagated crops. Elsevier, Amsterdam. In press.

Broertjes and Harten, van. 1985. Euphytica 34: 93–95

Broertjes et al. 1976. Euphytica 25: 11–19

Brown and Sommer. 1975. An Atlas of gymnosperms cultured in vitro: 1924–1974. Georgia Forest Res. Counc. Macon Georgia: 1–271

Bruijne, de and Debergh. 1974. Meded. Fac. Landbouwwet. Rijksuniv. Gent 39: 210–215

Brutti and Caso. 1986. Int. Congr. Plant Tissue Cell Culture Abstr. 6: 401

Bry, de. 1986. Newsletter I.A.P.T.C. 49: 2–22

Bulard and Dégivry. 1965. Phyton 22: 55–60

Bulard and Monin. 1963. Phyton 20: 115–125

Burgess. 1983. Intern. Rev. Cytol. Suppl. 16: 55–77

Burström. 1957. Physiol. Plant. 10: 741–751

Bush et al. 1974. HortScience 9: 270

Bush et al. 1976. Am. J. Bot. 63: 729–737

Butcher. 1977. In: Reinert and Bajaj. 1977: 668–693

Butcher and Ingram. 1976. Plant tissue culture. E. Arnold, London: 1–68

Butenko. 1968. Plant tissue culture and plant morphogenesis. Israel Program for Scientific Translations. Jerusalem: 1–291

Button. 1977. Outlook on Agric. 9: 155–159

Button and Bornman. 1971a. J. South Afr. Bot. 37: 127–134

324

Button and Bornman. 1971b. Citrus Grower and Sub-Trop. Fruit J. 453: 11–14

Button and Bornman. 1971c. Citrus Grower and Sub-Trop. Fruit J. 451: 24–25

Button and Kochba. 1977. In: Reinert en Bajaj. 1977: 70–92

Cameron et al. 1986. HortSci. 21: 774

Capite, de. 1955. Am. J. Bot. 42: 869–873

Carantino. 1983. Biofutur 13: 47–49

Carew and Staba. 1965. Lloydia 28: 1–26

Carlson. 1970. Science 168: 487–489

Carlson. 1973. Science 180: 1366–1368

Carlson and Chaleff. 1974. In: Ledoux. 1974: 245–261

Carlson et al. 1972. Proc. Nat. Acad. Sci. U.S.A. 69: 2292–2294

Cassells. 1985. In: Schäfer-Menuhr. 1985: 111–120

Cassells and Plunkett. 1986. In: Withers and Alderson. 1986: 105–111

Cassells et al. 1980. In: Ingram and Helgeson. 1980: 125–130

Champagnat. 1977. In: Gautheret. 1977: 238–253

Champagnat and Morel. 1969. Bull. Soc. Bot. France Mém. 116: 111–132

Champagnat and Morel. 1972. C.R. Acad. Sci. 274: 3379–3380

Champagnat et al. 1966. Rev. Gen. Bot. 73: 706–746

Chandler et al. 1972. Can. J. Bot. 50: 2265–2270

Chandra and Hildebrandt. 1967. Virology 31: 414–421

Chaussat en Bigot. 1980. La multiplication végétative des plantes supérieures. Gauthier-Villars, Paris, France: 1–277

Chaussat et al. 1986. Physiol. Vég. 24: 283–291

Chavadej and Becker. 1984. Plant Cell Tissue Organ Culture 3: 265–272

Cheema and Sharma. 1983. In: Sen and Giles. 1983: 309–317

Cheng and Voqui. 1977. Science 189: 306–307

Cheng-Haw et al. 1976–1978. Malayan Ochid Rev. 13: 60–65

Chiang-Shiang et al. 1976–1978. Malayan Orchid Rev. 13: 43–55

Chilton et al. 1977. Cell 11: 263–271

Churchill et al. 1971a. Am. Orchid. Soc. Bull. 40: 109–113

Churchill et al. 1971b. Die Orchidee 22: 147–151

Churchill et al. 1973. New Phytol. 72: 161–166

Cocking. 1960. Nature 187: 927–929

Cocking. 1977. Span 20: 5–7

Cocking. 1977a. Proc. 8th Congr. Eucarpia Interspecific Hybrid. Plant Breed.: 229–235

Cocking. 1978. In: Anonymous. 1978: 255–263

Cohen. 1986. Int. Congr. Plant Tissue Cell Culture Abstr. 6: 305

Coleman et al. 1980. Physiol. Plant. 48: 519–525

Collins et al. 1984. In: Gustafson. 1984: 323–383

Compton and Preece. 1986. Newl. I.A.P.T.C. 50: 9–18

Conger. 1981. Cloning agricultural plants via in vitro techniques. CRC Press, Boca Raton, Florida: 1–273

Conner and Thomas. 1981. Comb. Proc. Intern. Plant Prop. Soc. 31: 342–357

Constabel et al. 1974. Planta Medica 25: 158–165

Constabel and Eilert. 1986. Newsl. I.A.P.T.C. 50: 2–8

Cooke. 1977. HortScience 12(4): 339

Coumans et al. 1979. Acta Hort. 91: 287–293

Cox et al. 1945. Plant Physiol. 20: 289

Cox et al. 1960. Nature 185: 403–404

Creze. 1985. C.R. Séances Acad. Agric. France 71: 572–576

Crisp and Walkey. 1974. Euphytica 23: 305–313
Cronauer and Krikorian. 1986. Int. Congr. Plant Tissue Cell Culture Abstr. 6: 179

Dale. 1976. Rep. Welsh Plant Breed. Station: 101–115
Dale and Webb. 1985. In: Bright and Jones. 1985: 79–96
D'Amato. 1977. In: Reinert and Bajaj. 1977: 343–357
D'Amato. 1978. In: Thorpe. 1978: 287–296
D'Amato. 1985. Critical Rev. Plant Sci. 3: 73–112
Damiano et al. 1986. Hort. Sci. 21: 684
Datta et al. 1983. Plant Cell Tissue Organ Culture 2: 15–20
Daub. 1986. Ann. Rev. Phytopath. 24: 159–186
Davey and Kumar. 1983. Intern. Rev. Cytol. Suppl. 16: 219–299
David. 1982. In: Bonga and Durzan. 1982: 72–108
Debergh. 1974. Antheren- en pollenkultuur voor het bekomen van haploide embryo's van
    Lycopersicon spp. Diss. Rijksuniv. Gent, Belgium: 1–137
Debergh. 1978. Verh. Kon. Acad. Wet. Belg. 40: 1–70
Debergh. 1982. Nieuwe aspecten bij de weefselteelt van planten. Thesis Rijksuniv. Gent,
    Belgium: 1–126
Debergh. 1983. Physiol. Plant. 59: 270–276.
Debergh. 1986. Int. Congr. Plant Tissue Cell Culture Abstr. 6: 430
Debergh and Maene. 1977. Acta Hortic. 78: 449–454
Debergh and Maene. 1981. Scientia Hortic. 14: 335–345
Debergh et al. 1981. Physiol. Plant. 53: 181–187
Degani and Smith. 1986. Int. Congr. Plant Tissue Cell Culture Abstr. 6: 411
Dégivry. 1966. Bull. Sci. Bourgogne 24: 57–87
Deleplanque et al. 1985. Proc. 5th Intern. Conf. Robot Vision Sensory Control. Kempston,
    Bedford: 305–314
Demarly. 1986. Zaadbelangen 40: 41–42
Desjardins et al. 1986. HortSci. 21: 735
Dhawan et al. 1986. Int. Congr. Plant Tissue Cell Culture Abstr. 6: 113
Dietrich. 1924. Flora 17: 379–417
Dixon. 1985. Plant cell culture, a practical approach. IRL Press, Oxford: 1–236
Dodds. 1983. Tissue culture of trees. Croom Helm, London: 1–147
Dodds. 1985. Plant genetic engineering. Cambridge Univ. Press, Cambridge: 1–312
Dolfus and Nicolas-Prat. 1969. C.R. Acad. Sci, 268: 501–503
Dommergues and Gillot. 1973. Ann. Amélior. Plantes 23: 89–93
Dons and Bouwer. 1986. In: Anonymous. 1986: 497–504
Dons and Jong, de. 1986. Vakbl. Bloem. 41(43): 80–83
Doré. 1980. Application de la culture in vitro à l'amélioration des plantes potagères. CNRA,
    INRA, Versailles: 1–210
Dougall. 1979. In: Sharp et al. 1979: 727–743
Dougall and Wetherell. 1974. Cryobiology 11: 410–415
Draper et al. 1986. In: Yeoman. 1986: 271–301
Drira et al. 1986. Int. Congr. Plant Tissue Cell Culture Abstr. 6: 181
Dudits et al. 1976. Can. J. Bot. 54: 1063–1067
Dudits et al. 1976a. Cell genetics in higher plants. Akademiae Kiado, Budapest: 1–251
Dulieu. 1972. Phytomorph. 22: 283–296
Dulieu and Barbier. 1982. In: Earle and Demarly. 1982: 211–227
Dunwell. 1985. In: Bright and Jones. 1985: 1–44

Earle and Demarly. 1982. Variability in plants regenerated from tissue culture. Praeger, New York: 1–392

Earle and Langhaus. 1974a. J. Am. Soc. Hort. Sci. 99: 128–131

Earle and Langhans. 1974b. J. Am. Soc. Hort. Sci. 99: 352–358

Economou. 1986. Int. Congr. Plant Tissue Cell Culture Abstr. 6: 431

Economou and Read. 1986. HortSci. 21: 761

Elliot. 1970. Planta 95: 183–186

Englis and Hanahan. 1945. J. Am. Chem. Soc. 67: 51–54

Eriksson and Jonassen. 1969. Planta 89: 85–89

Ernst. 1975. Am. Orchid Soc. Bull. 44: 12–18

Evans and Bravo. 1983. In: Evans et al. 1983: 124–176

Evans and Bravo. 1983a. Intern. Rev. Cytol. Suppl. 16: 33–53

Evans and Bravo. 1986. In: Zimmerman. 1986: 73–94

Evans and Sharp. 1986. In: Evans et al. 1986: 97–132

Evans et al. 1981. In: Thorpe. 1981: 45–113

Evans et al. 1983. Handbook of plant cell culture. Vol. 1. Techniques for propagation and breeding. MacMillan Pulb. Co., New York: 1–970

Evans et al. 1984. Am. J. Bot. 71: 759–774

Evans et al. 1986. Handbook of plant cell culture. Vol. 4. Techniques and applications. MacMillan Publ. Co., New York: 1–698

Evers. 1984. Growth and morphogenesis of shoot initials of Douglas fir, *Pseudotsuga mensiesii* (Mirb.) Franco, in vitro. Diss. Agric. Univ. Wageningen, the Netherlands. Article 1–6

Fabbri et al. 1986. Sci. Hortic. 28: 331–337

Fast. 1971. Die Orchidee 22: 189–192

Fast. 1980. Orchideen Kultur. Verlag E. Ulmer, Stuttgart: 1–460

Feng and Link. 1970. Plant Cell Physiol. 11: 589–598

Fiechter. 1985. Plant cell culture. Adv. Biochem. Engineering Biotechn., Springer Verlag, Berlin: 1–140

Fillatti et al. 1986. HortSci. 21: 773

Flamee. 1978. Am. Orchid Soc. Bull. 47: 419–423

Flemion. 1948. Contr. Boyce Thomson Instit. 15: 229–241

Flick and Evans. 1983a. In: Randall et al. 1983: 200–210

Flick et al. 1983. In: Evans et al. 1983: 13–81

Fonnesbech. 1972. Physiol. Plant. 27: 310–316

Fossard, de. 1976. Tissue culture for plant propagators. Univ. New England Printery, Armidale, Australia: 1–409

Fossard, de. 1977. The horizons of tissue culture propagation Seminar NSW ASS. Nurserymen. Univ. Sydney: 1–147

Fouret et al. 1986. Int. Congr. Plant Tissue Cell Culture Abstr. 6: 371

Fowler. 1984. In: Vasil. 1984: 167–174

Fowler. 1986. In: Yeoman. 1986: 202–227

Frett and Smagula. 1983. Can. J. Plant. Sci. 63: 467–472

Fridlund. 1980. In: Anonymous. 1980a: 86–92

Fujimura and Komamine. 1984. In: Vasil. 1984: 159–166

Fujiwara (ed). 1982. Plant tissue culture 1982. Proc. 5th Int. Congr. Plant Tissue Cell Culture, Tokyo, Japan: 1–839

Galston. 1947. Am. J. Bot. 34: 599–600

Galston. 1948. Am. J. Bot. 35: 281–287

327

Galun and Aviv. 1984. In: Lange et al. 1984: 228–238
Galzy. 1969a. Ann. Phytopathol. Physiol. 1: 149–166
Galzy. 1969b. Vitis 8: 191–205
Gamborg and Wetter. 1975. Plant tissue culture methods. Publ. NRC-CNRC. Ottawa, Canada: 1–110
Gamborg et al. 1968. Exp. Cell Res. 50: 151–158
Gamborg et al. 1974. Can. J. Genet. Cytol. 16: 737–750
Gamborg et al. 1976. In Vitro 12: 473–478
Gas et al. 1971. C.R. Acad. Sci. 272: 407–410
Gathercole et al. 1976. Physiol. Plant. 37: 213–217
Gautheret. 1934. C.R. Acad. Sci. 198: 2195–2196
Gautheret. 1939. C.R. Acad. Sci. 208: 118–120
Gautheret. 1940. C.R. Acad. Sci. 210: 632–634
Gautheret. 1959. La culture des tissues végétaux. Masson Cie. Paris: 1–863
Gautheret. 1964. Rev. Cytol. Biol. Vég. 32: 99–220
Gautheret. 1969. Am. J. Bot. 56: 702–717
Gautheret. 1977. La culture des tissues et des cellules des végétaux. Masson, Paris: 1–261
Gautheret. 1983. Bot. Mag. Tokyo 96: 393–410
Gebhard et al. 1983. Mitt. Klosterneuburg 33: 24–30
Geier. 1983. Acta Hortic. 131: 329–337
Gengenbach and Green. 1975. Crop. Sci. 15: 645–649
George and Sherrington. 1984. Plant propagation by tissue culture. Eversley, United Kingdom: 1–709
George et al. 1983. Plant Cell Rep. 2: 189–191
Gheysen et al. 1985. In: Hohn and Dennis. 1985: 11–47
Gibson. 1967. Austr. J. Biol. Sci. 20: 837–842
Gleba and Sytnik. 1984. Protoplasts fusion, genetic engineering in higher plants. Springer Verlag, Berlin: 1–220
Goldy and Lyrene. 1984. J. Am. Soc. Hort. Sci. 109: 336–338
Gonzales and Widholm. 1985. In: Dixon. 1985: 67–78
Gorter. 1965. J. Hort. Sci. 40: 177–179
Gray. 1981. Hort. Rev. 3: 1–27
Gray. 1986. HortSci. 21: 820
Greenwood and Berlyn. 1973. Am. J. Bot. 60: 42–47
Griesbach. 1981. Plant Cell Tissue Organ Culture 1: 103–107
Griffis et al. 1983. Comb. Proc. Intern. Plant Prop. Soc. 33: 618–622
Groot, de et al. 1982. Genetische manipulatie in dienst van de landbouw. Nat. Raad voor Landbouwkundig Onderzoek, The Hague, the Neterhlands. Studierapport 14a: 1–63
Grout. 1975. Plant Science Lett. 5: 401–405
Grout and Aston. 1977. Hort. Res. 17: 1–7
Grout and Aston. 1978. Hort. Res. 17: 65–71
Grout and Crisp. 1977. Acta Hort. 78: 289–296
Grout et al. 1986. Int. Congr. Plant Tissue Cell Culture Abstr. 6: 148
Grunewaldt. 1983. Acta Hortic. 131: 339–343
Guha and Maheswari. 1964. Nature 204: 479
Guha and Maheswari. 1966. Nature 212: 97–98
Gunn and Day. 1986. In: Withers and Alderson. 1986: 313–336
Gupton. 1986. HortSci. 21: 734
Gustafson. 1984. Gene manipulation in plant improvement. Plenum Press, New York. 1984: 1–668

Haberlandt. 1902. Sitzungsber. Akad. Wiss. Wien Math. Nat. Klasse 111(1): 69–92
Haccius. 1978. Phytomorph. 28: 74–81
Haccius and Hausner. 1975. Planta Medica Suppl.: 35–41
Hackett. 1970. J. Am. Soc. Hort. Sci. 95: 398–402
Hackett. 1985. Hort. Rev. 7: 109–155
Hackett and Anderson. 1967. Proc. Am. Soc. Hort. Sci. 90: 365–369
Hadley. 1975. In: Sanders et al. 1975: 335–351
Hadley. 1982. In: Arditti. 1982: 83–118
Hakkaart and Balen, van. 1980. Vakbl. Bloem. 35(11): 34–37
Hakkaart and Versluijs. 1984. Vakbl. Bloem. 39(49): 49–50
Hakkaart et al. 1980. Vakbl. Bloem. 35(2): 24–25
Hakkaart et al. 1983. Vakbl. Bloem. 38(33): 38–39
Hall. 1948. Proc. Am. Soc. Hort. Sci. 52: 343–346
Halperin. 1969. Ann. Rev. Plant Physiol. 20: 395–418
Halquist et al. 1983. In Vitro 19: 248
Hammerschlag. 1982. J. Am. Soc. Hort. Sci. 107: 44–47
Hännig. 1904. Bot. Zeitung. 62: 45–80
Hannings and Langhaus. 1974. HortScience 9: 271
Hansen and Lane. 1985. Plant Disease 69: 134–135
Hanson and Edelman. 1972. Planta 102: 11–25
Harberd. 1969. Euphytica 18: 425–429
Harkins and Galbraith. 1984. Physiol. Plant. 60: 43–52
Harley. 1969. The biology of mycorrhiza. L. Hill, London, second edition: 1–334
Harms. 1983. Quart. Rev. Biol. 58: 325–353
Harms. 1984. In: Vasil. 1984: 213–220
Harms et al. 1976. Z. Pflanzenzücht. 77: 347–351
Harten, van et al. 1981. Euphytica 30: 1–8
Heberle-Bors. 1985. Theor. Appl. Gen. 71: 361–374
Heble. 1985. In: Neumann et al. 1985: 281–289
Heide. 1965. Physiol. Plant. 18: 891–920
Hein and Schieder. 1984. In: Lange et al. 1984: 248–250
Heinz. 1973. Induced mutations in vegetatively propagated plants: 53–59
Heller. 1953. Ann. Sci. Bot. Vég. Paris 14: 1–223
Henke et al. 1985. Tissue culture in forestry and agriculture. Basic Life Sciences vol. 32.
    Plenum Press, New York: 1–390
Henshaw. 1974. Abstr. 3rd Int. Congr. Plant Tissue and Cell Culture, Leicester: 212
Henshaw. 1978. Biol. J. Linnean Soc. 10: 434
Henshaw. 1982. Proc. 5th Int. Congr. Plant Tissue Cell Culture 5: 789–792
Henshaw and O'Hara. 1984. In: Mantell and Smith. 1984: 219–238
Herman and Haas. 1978. Z. Pflanzenphysiol. 89: 467–470
Hermsen. 1977. Vakblad Biologen 57: 86–91
Hermsen. 1984. Iowa State J. Res. 58: 449–460
Hermsen and Ramanna. 1981. Phil. Trans. Roy. Soc. London B292: 499–507
Hermsen et al. 1981. Euphytica 30: 239–246
Hernalsteens et al. 1980. Nature 287: 654–656
Hess and Wagenr. 1974. Z. Pflanzenphysiol. 72: 466–468
Hildebrandt. 1971. Coll. Int. C.N.R.S. 193: 71–93
Hildebrandt. 1977: In: Reinert and Baja. 1977: 581–597
Hiroaka. 1986. Int. Congr. Plant Tissue Cell Culture Abstr. 6: 37
Hohn and Dennis. 1985. Genetic flux in plants. Springer Verlag, Vienna: 1–253
Homès et al. 1972. Bull. Soc. Roy. Belg. 106: 123–128

Hondelmann and Wilberg. 1973. Z. Pflanzenzüchtung 69: 19–24
Hooykaas–van Slogteren et al. 1984. Nature 311: 763–764
Horak et al. 1975. Ann. Bot. 39: 571–577
Horgan. 1986. Int. Congr. Plant Tissue Cell Culture Abstr. 6: 295
Horsch et al. 1985. Science 227: 1229–1231
Houten, ten et al. 1968. Neth. J. Plant Path. 74: 17–24
Howland and Hart. 1977. In: Reinert and Bajaj. 1977: 731–756
Hu and Wang. 1983. In: Evans et al. 1983: 177–227
Hu and Wang. 1986. In: Evans et al. 1986: 43–96
Huang and Millikan. 1980. HortScience 15: 741–743
Hughes. 1981. In: Conger. 1981: 5–50
Hughes et al. (eds). 1978. Propagation of higher plants through tissue culture. U.S. Dept. of
    Energy, Tech. Comm. Center: 1–305
Hunter. 1979. J. Hort. Sci. 54: 111–114
Hunter et al. 1986. Int. Congr. Plant Tissue Cell Culture Abstr. 6: 410
Hussey. 1986. In: Withers and Alderson. 1986: 69–84
Hussey and Falavigna. 1980. J. Exp. Bot. 31: 1675–1686
Huxter et al. 1981. Physiol. Plant. 53: 319–326
Hyner and Arditti. 1973. Am. J. Bot. 60: 829–835

Ingle. 1982. Span 25: 50–53
Ingram. 1973. In: Street. 1973: 392–421
Ingram and Helgeson. 1980. Tissue culture methods for plant pathologists. Blackwell Scient.
    Publ., Oxford: 1–272
Ingram and Macdonald. 1986. In: Anonymous. 1986: 241–258
Intuwong and Sagawa. 1973. Am. Orchid Soc. Bull. 42: 209–215
Iwai and Kishi. 1986. Int. Congr. Plant Tissue Cell Culture Abstr. 6: 82

Jacob et al. 1980. Meded. Fac. Landb. Wet. Rijksuniv. Gent 45: 335–343
Jacquiot. 1951. C.R. Acad. Sci. 233: 815–817
Jacquiot. 1964. Ann. Sci. Forest. Nancy 21: 315–474
Jacquiot. 1966. J. Inst. Wood Sci. 16: 22–34
Jaiswal. 1986. Int. Congr. Plant Tissue Cell Culture Abstr. 6: 163
Jansen. 1982. In: Fujiwara. 1982: 880–888
Jensen. 1974. Production of monoploids in barley. A progress report. In: Polyploidy and
    induced mutations in plant breeding, Int: Atom. Energy Agency PL 503/24: 167–169
Jensen. 1976. Newsl. I.A.P.T.C. 18: 2–7
Jensen. 1977. In: Reinert en Bajaj. 1977: 299–340
Johansson. 1983. Physiol. Plant. 59: 397–403
Johansson and Eriksson. 1982. Proc. 5th Int. Plant Tissue Cell Culture 5: 543–544
Johnson. 1980. HortScience 15: 605–606
Johri and Bhojwani. 1977. In: Reinert en Bajaj. 1977: 398–411
Johri et al. 1982. Experimental embryology of vascular plants. Springer Verlag, Berlin: 1–
    273
Jonard. 1986. In: Bajaj. 1986: 31–48
Jonard et al. 1983. Scientia Hortic. 20: 147–159
Jones. 1979. Scientia Hortic. 30: 44–48
Jones. 1983. Biologist 30: 181–188
Jones. 1986. In: Zimmerman et al. 1986: 175–182
Jones and Hopgood. 1979. J. Hort. Sci. 54: 63–66

Kadkade and Jobson. 1978. Plant Sci. Lett. 13: 67–73
Kahn. 1976. Plant Disease Rep. 60: 459–461
Kahn. 1986. In: Zimmerman et al. 1986: 147–164
Kameya. 1975. Jap. J. Gen. 50: 417–420
Kameya. 1984. In: Vasil. 1984: 423–427
Kanta. 1960. Nature 188: 683–684
Kanta et al. 1962. Nature 194: 1214–1217
Kao and Michayluk. 1975. Planta 126: 105–110
Karp et al. 1984. Plant Cell Tissue Organ Culture 3: 363–373
Karper and Pierik. 1981. Vakbl. Bloem. 36(44): 44–45
Kartha. 1981. In: Thorpe. 1981: 181–211
Kartha. 1985. Cryopreservation of plant cells and organs. CRC Press Inc., Boca Raton, U.S.A.: 1–276
Kartha. 1986. In: Withers and Alderson. 1986: 219–238
Kartha and Gamborg. 1975. Phytopathology 65: 826–828
Kartha et al. 1974. Abstr. 3rd Int. Congr. Plant Tissue and Cell Culture, Leicester: 117
Kasha. 1974. Haploids in higher plants. Univ. Guelph, Canada: 1–421
Kasha and Kao. 1970. Nature 225: 874–876
Kasperbauer and Collins. 1972. Crop. Sci. 12: 98–101
Keller. 1986. Int. Congr. Plant Tissue Cell Culture Abstr. 6: 120
Kester. 1953. Hilgardia 22: 335–365
Kinderen, van der. 1982. Orchideeën 44: 96–97
King. 1984. In: Lange et al. 1984: 199–214
King and Street. 1973. In: Street. 1973: 269–337
Kitto and Janick. 1985. J. Am. Soc. Hort. Sci. 110: 277–282
Kitto and Janick. 1985a. J. Am. Soc. Hort. Sci. 110: 283–286
Klein and Bopp. 1971. Nature 230: 474
Knauss. 1976. Florida State Hort. Soc. 89: 293–296
Knauss and Miller. 1978. In Vitro 14: 754–756
Knop. 1884. Landw. Vers. Stn. 30: 292–294
Knudson. 1922. Bot. Gaz. 73: 1–25
Knudson. 1946. Am. Orchid Soc. Bull. 15: 214–217
Koblitz. 1969. Biol. Rundschau 7: 241–249
Koblitz. 1972. Zell- und Gewebezüchtung bei Pflanzen. G. Fishcer Verlag, Jena: 1–84
Kochar et al. 1971. J. Hered. 62: 59–61
Kohlenbach. 1977. In: Barz et al. 1977: 355–366
Kohlenbach. 1978. In: Thorpe. 1978: 59–66
Kohlenbach. 1985. In: Schäfer-Menuhr. 1985: 101–109
Konar and Nataraja. 1969. Acta Bot. Neerl. 18: 680–699
Kott and Kasha. 1985. In: Bright and Jones. 1985: 45–78
Kovoor. 1962. C.R. Acad. Sci. 255: 1991–1993
Krens et al. 1982. Nature 296: 72–74
Krusberg and Babineau. 1979. In: Sharp et al. 1979: 401–419
Kruyt. 1951. Vakblad Biologen 31: 181–193
Kruyt. 1952. Proc. Kon. Ned. Akad. Wetensch. C 55: 503–513
Kumar et al. 1986. Int. Congr. Plant Tissue Cell Culture Abstr. 6: 11
Kunneman-Kooij. 1984. Inventarisatie van de toepassingsmogelijkheden van weefselkweek voor de vegatatieve vermeerdering van houtige boomkwekerijgewassen. Literature study, Department of Horticulture, Agricultural Univ. Wageningen, the Netherlands: 1–64

Kuster. 1909. Ber. Deut. Bot. Ges. 27: 589–598
Kyte. 1983. Plants from test tubes. An introduction to micropropagation. Timber Press, Portland, Oregon: 1–132
Laibach. 1925. Z. Bot. 17: 417–459
Laibach. 1929. J. Hered. 20: 201–208
Lakso et al. 1986. Int. Congr. Plant Tissue Cell Culture Abstr. 6: 397
Lammerts. 1942. Am. J. Bot. 29: 166–171
Lammerts. 1946. Plants Gardens 2: 11
Lamotte and Lersten. 1972. Am. J. Bot. 59: 89–96
Langford and Wainwright. 1986. Int. Congr. Plant Tissue Cell Culture Abstr. 6: 433
Lange. 1969. Diss. Wageningen. Verslagen Landb. Onderz. 719: 1–162
Lange et al. 1984. Efficiency in plant breeding. Pudoc, Wageningen: 1–383
Larebeke et al. 1974. Nature 252: 169–170
Larkin. 1982. Plant Cell Tissue Organ Culture 1: 149–164
Larkin and Scowcorft. 1981. Theor. Appl. Genetics 60: 197–214
LaRue. 1936. Bull. Torrey Bot. Club 63: 365–382
Latta. 1971. Can. J. Bot. 49: 1253–1254
Ledoux. 1974. Genetic manipulations with plant materials. Plenum Press, New York: 1–601
Lee and Mowe. 1983. Singapore J. Primary Industries 11: 46–48
Leffring. 1968. Meded. Dir. Tuinbouw 31: 392–395
Lescure. 1970. Bull. Soc. Bot. France 117: 353–365
Letham. 1974. Physiol. Plant. 32: 66–70
Letouzé. 1974. Physiol. Vég. 12: 397–412
Letouzé and Beauchesne. 1969. C.R. Acad. Sci. 269: 1528–1531
Limasset and Cornuet. 1949. C.R. Acad. Sci. Paris 228: 1888–1890
Lindsey and Yeoman. 1985. In: Neumann et al. 1985: 304–315
Lindsey and Yeoman. 1986. In: Yeoman. 1986: 228–267
Linskens et al. 1983. Plantenveredeling en genen manipulatie. Pudoc, Wageningen, the Netherlands: 1–150
Litz. 1985. In: Henke et al. 1985: 179–193
Litz et al. 1985. Hort. Rev. 7: 157–200
Liu and Shang. 1974. Abstr. 3rd Int. Congr. Plant Tissue and Cell Culture, Leicester: 46
Lloyd and McCown. 1980. Comb. Proc. Int. Plant Prop. Soc. 30: 421–427
Long and Cassells. 1986. In: Withers and Anderson. 1986: 239–248
Loo. 1945. Am. J. Bot. 32: 13–17
Lucke. 1969. Die Orchidee 20: 270–271
Lucke. 1971. Die Orchidee 22: 146–147
Lyrene. 1984. Florida Agric. Res. 3: 23–25
Maat. 1980. Gewasbescherming 1980, 11: 13–21
Maat and Schuring. 1981. Vakblad Bloem. 36(10): 34–35
Mackenzie and Street. 1970. J. Exp. Bot. 21: 824–834
Maene and Debergh. 1983. Acta Hort. 131: 201–208
Maene and Debergh. 1985. Plant Cell Tissue Organ Culture 5: 23–33
Magnien and Nettancourt, de. 1985. Genetic engineering of plants and microorganisms important for agriculture. M. Nijhoff, Dordrecht, the Netherlands: 1–197
Maheshwari and Rangaswamy. 1958. Indian J. Hortic. 15: 275–281
Maheshwari and Rangaswamy. 1965. Adv. Bot. Res. 2: 219–321
Maheshwari et al. 1983. Newsletter I.A.P.T.C. 41: 2–9
Maliga. 1984. Ann. Rev. Plant Physiol. 35: 519–542

Mantell and Smith. 1984. Plant biotechnoloty. Cambridge University Press, Cambridge, England: 1-334

Marcotrigiano and Gouin. 1984. Ann. Bot. 54: 513-522

Marcotrigiano and Gouin. 1984a. Ann. Bot. 54: 503-511

Maretzki et al. 1971. Plant Physiol. 48: 521-525

Margara. 1969. Ann. Physiol. Vég. 11: 95-112

Margara. 1982. Bases de la multiplication végétative, les méristèmes et l'organogenèse. Publ. Inst. Nat. Rech. Agron. Paris: 1-262

Marshall. 1977. A survey of laboratories used for plant tissue culture propagation in California. Misc. Publ. Ornamental Hort. California Polytechnic State Univ., San Luis Obispo: 1-106

Martin-Tanguy et al. 1976. C.R. Acad. Sci. 282: 2231-2234

Marton et al. 1979. Nature 277: 129-131

Mathes. 1964. Phyton 21: 137-141

Matsubara et al. 1974. J. Jap. Soc. Hort. Sci. 43: 63-68

Mayer. 1956. Planta 47: 401-446

Mbanaso and Roscoe. 1982. Plant Sci. Lett. 25: 61-66

McComb. 1978. In: Anonymous. 1978a: 167-180

McCown. 1986. In: Zimmerman et al. 1986: 53-60

Meins. 1983. Ann. Rev. Plant Physiol. 34: 327-346

Melchers and Labib. 1974. Molec. Gen. Genet. 135: 277-294

Melchers et al. 1978. Carlsberg Res. Comm. 43: 203-218

Melé et al. 1982. Proc. 5th Int. Congr. Plant Tissue Cell Culture 5: 69-70

Mellor and Stace-Smith. 1969. Can. J. Bot. 47: 1617-1621

Menczel. 1983. What's new in Plant Physiology 14: 9-11

Meredith. 1984. In: Gustafson. 1984: 503-528

Merrick and Collin. 1980. In: Doré. 1980: 121-125

Mestre and Benbadis. 1985. Bull. Soc. Bot. France 132: 1-171

Meyer. 1983. Comb. Proc. Int. Plant Prop. Soc. 33: 402-407

Meyer and Kernsh. 1986. Int. Congr. Plant Tissue Cell Culture Abstr. 6: 149

Meynet and Sibi. 1984. Z. Pflanzenzüchtung 93: 78-85

Miller. 1985. In: Dixon. 1985: 215-229

Miller and Maxwell. 1983. In: Evans et al. 1983: 853-879

Miller and Skoog. 1953. Am. J. Bot. 40: 768-773

Miller et al. 1955. J. Am. Chem. Soc. 77: 2662-2663

Misson et al. 1983. Meded. Fac. Landbouwwet. Rijksuniv. Gent 48: 1151-1157

Mix. 1984. In: Lange et al. 1984: 194-195

Mohan Ram et al. 1975. Form, structure and function in plants. Sarita Prakashan, India: 1-457

Mok et al. 1986. HortSci. 21: 796

Moncousin. 1979. Revue Hort. Suisse 52: 361-363

Moncousin. 1982. Proc. 5th Int. Congr. Plant Tissue Cell Culture 5: 147-148

Monnier. 1976. Rev. Cytol. Biol. Vég. 39: 1-120

Monnier. 1980. Bull. Soc. Bot. France 127: 59-70

Montant. 1957. C.R. Séanc. Soc. Biol. 151: 391-392

Morandi et al. 1979. Ann. Amélior. Plantes 29: 623-630

Morel. 1960. Am. Orchid Soc. Bull. 29: 495-497

Morel. 1964. Rev. Hort. Ann. Soc. Nat. Hort. France 136: 733-740

Morel. 1965. Gartenwelt 65: 350-352

Morel. 1974. In: Withner. 1974: 169-222

Morel and Martin. 1952. C.R. Acad. Sci. 235: 1324-1325

Morel and Martin. 1955. C.R. Acad. Agric. France 41: 472–475
Mori et al. 1982. Proc. 5th Int. Congr. Plant Tissue Cell Culture 5: 803–804
Morris et al. 1985. In: Dixon. 1985: 127–167
Mosella. 1979. Diss. Université Sciences Techn. Languedoc: 1–202
Mosella et al. 1979. C.R. Acad. Sci. Paris 289: 505–508
Mott. 1981. In: Conger. 1981: 217–254
Muir et al. 1954. Science 119: 87–878
Muir et al. 1958. Am. J. Bot. 45: 589–597
Murakishi and Carlson. 1979. In Vitro 15: 192 (Abstr. 114)
Murashige. 1961. Science 134: 280
Murashige. 1964. Physiol. Plant. 17: 636–643
Murashige. 1974. Ann. Rev. Plant Physiol. 25: 135–166
Murashige and Nakano. 1966. J. Hered. 57: 115–118
Murashige and Skoog. 1962. Physiol. Plant. 15: 473–497
Murashige et al. 1972. HortScience 7: 118–119
Murashige et al. 1974. HortScience 9: 175–180

Nag and Street. 1975a. Physiol. Plant. 34: 254–260
Nag and Street. 1975b. Physiol. Plant. 34: 261–265
Nakamura. 1982. New Phytol. 90: 701–715
Narayanaswamy and Norstog. 1964. Bot. Rev. 30: 587–628
Navarro et al. 1975. J. Am. Soc. Hort. Sci. 100: 471–479
Németh. 1979. Z. Pflanzenphysiol. 95: 389–396
Németh. 1986. In: Bajaj, 1986: 49–64
Neumann et al. 1985. Primary and secondary metabolism of plant cell cultures. Springer
    Verlag, Berlin: 1–377
Nickell. 1951. Proc. Am. Soc. Hort. Sci. 57: 401–405
Nickell. 1973. Hawaiian Plant. Record 58: 293–314
Niimi. 1970. J. Jap. Soc. Hort. Sci. 34: 346–352
Nitsch. 1968. Ann. Sci. Nat. Bot. Biol. Vég. 9: 1–92
Nitsch. 1969. Phytomorphology 19: 389–404
Nitsch. 1972. Z. Pflanzenzüchtung 67: 3–18
Nitsch. 1977. In: Reinert and Bajaj. 1977: 268–278
Nitsch. 1981. In: Thorpe. 1981: 241–252
Nitsch et al. 1967. Phytomorphology 17: 446–453
Nitsch et al. 1969a. C.R. Acad. Sci. 269: 1257–1278
Nitsch et al. 1969b. C.R. Acad. Sci. 269: 1650–1652
Nitzsche. 1983. In: Evans et al. 1983: 782–805
Nobécourt. 1939. C.R. Séance Soc. Biol. 130: 1270–1271
Noiton et al. 1986. HortSci. 21: 706
Norris et al. 1983. Science 220: 75–76
Norstog. 1975. In: Mohan Ram e tal. 1975: 197–201
Norton and Norton. 1986. Plant Cell Tissue Organ Culture 5: 187–197
Norton and Norton. 1986a. HortSci. 21: 699
Norton and Norton. 1986b. Int. Congr. Plant Tissue Cell Culture Abstr. 6: 434

Ooms et al. 1983. Theor. Appl. Genet. 66: 169–172
Opatrny and Landa. 1974. Biol. Plant. 16: 312–315
Orton. 1983. Plant Mol. Biol. Rep. 1: 67–76
Orton. 1984. In: Gustafson. 1984: 427–468
Os, van. 1964. Neth. J. Plant Pathology 70: 18–26
Overbeek, van et al. 1941. Science 94: 350–351

334

Palmer and Barker. 1973. Ann. Bot. 37: 85–93
Parliman. 1986. In: Zimmerman et al. 1986: 271–282
Paszkowski et al. 1984. EMBO J. 3: 2717–2722
Paulet. 1965. Rev. Gen. Bot. 72: 697–792
Peck and Cumming. 1986. Plant Cell Tissue Organ Culture 6: 9–14
Pelletier and Chupeau. 1984. Physiol. Vég. 22: 377–399
Pelletier and Delise. 1969. Ann. Amélior. Plantes 19: 353–355
Pelletier et al. 1983. Mol. Gen. Genet. 191: 244–250
Pennell. 1982. Introduction to the micropropagation of horticultural crops. Min. of Agric.
  United Kingdom 2413: 1–46
Pennell. 1985. Grower 103(8, Suppl.): 98–101
Perry and Lyrene. 1984. J. Am. Soc. Hort. Sci. 109: 4–6
Phatak. 1974. Seed Sci. Technol. 2: 3–155
Pierik. 1966. Naturwiss. 53: 387
Pierik. 1967. Meded. Landbouwhogeschool Wageningen 67(6): 1–71
Pierik. 1967a. Z. Pflanzenphysiol. 66: 141–152
Pierik. 1969. Neth. J. Agric. Sci. 17: 203–208
Pierik. 1972. Z. Pflanzenphysiol. 56: 343–351
Pierik. 1975. Plantenteelt in kweekbuizen. Thieme, Zutphen, the Netherlands: 1–164
Pierik. 1975c. Neth. J. Agric. Sci. 23: 299–302
Pierik. 1975d. Acta Hort. 54: 71–82
Pierik. 1976. Neth. J. Agric. Sci. 24: 98–104
Pierik. 1976a. Physiol. Plant. 37: 80–82
Pierik. 1976b. Groenten en Fruit 32(23): 15–17
Pierik. 1979. In vitro culture of higher plants. Bibliography. Ponsen en Looyen, Wageningen,
  the Netherlands: 1–149
Pierik. 1984. Vakbl. Bloem. 39(24): 44–45
Pierik. 1986. Vakbl. Bloem. 41(17): 68–69
Piereik and Ippel. 1977. Acta Hort. 78: 197–202
Pierik and Post. 1975. Scientia Hort. 3: 293–297
Pierik and Ruibing. 1973. Neth. J. Agric. Sci. 21: 129–138
Pierik and Segers. 1973. Z. Pflanzenphysiol. 69: 204–212
Pierik and Sprenkels. 1984. Vakbl. Bloem. 39(21): 44–45
Pierik and Steegmans. 1972. Z. Pflanzenphysiol. 68: 228–234
Pierik and Steegmans. 1975a. Scientia Hort. 3: 1–20
Pierik and Steegmans. 1975b. Neth. J. Agric. Sci. 23: 334–337
Pierik and Steegmans. 1975c. Vakblad Bloem. 30(24): 18–21
Pierik and Steegmans. 1975d. Physiol. Plant. 34: 14–17
Pierik and Steegmans. 1976. Scientia Hort. 4: 291–292
Pierik and Steegmans. 1976a. Neth. J. Agric. Sci. 24: 274–277
Pierik and Steegmans. 1983a. Scientia Hort. 21: 267–272
Pierik and Steegmans. 1984. Neth. J. Agric. Sci. 32: 101–106
Pierik and Steegmans. 1986. Vakbl. Bloem. 41(32): 53
Pierik and Steegmans. 1986a. Neth. J. Agric. Sci. 34: 217–223
Pierik and Tetteroo. 1986. Vakbl. Bloem. 41(46): 32–33
Pierik and Woets. 1971. Acta Hort. 23(2): 423–428
Pierik et al. 1973. Scientia Hort. 1: 117–119
Pierik et al. 1974. Vakbl. Bloem. 29(39): 18–21
Pierik et al. 1974a. Scientia Hort. 2: 193–198
Pierik et al. 1975. Vakbl. Bloem. 30(27): 17
Piereik et al. 1975a. Scientia Hort. 3: 351–357

Pierik et al. 1979. Vakbl. Bloem. 34(25): 36–37
Pierik et al. 1979a. Neth. J. Agric. Sci. 27: 221–226
Pierik et al. 1982. Neth. J. Agric. Sci. 30: 341–346
Pierik et al. 1982a. Vakbl. Bloem. 37(52): 38–40
Pierik et al. 1983. Orchideeën 45: 91–95
Pierik et al. 1983a. Acta Hort. 147: 179–186
Pierik et al. 1984. Vakbl. Bloem. 39(3): 33
Pierik et al. 1985. Bloembollencultuur 96(45): 8–10
Pierik et al. 1986. Int. Congr. Plant Tissue Cell Culture Abstr. 6: 434
Pierson et al. 1983. Protoplasma 115: 208–216
Pilet and Roland. 1971. Cytobiologie 4: 41–61
Poisson and Maia. 1974. Revue Hort. 2325: 33–36
Potrykus et al. 1983. Protoplasts 1983. Lecture Proceedings. Birkhäuser Verlag, Basel:
      1–267
Power and Chapman. 1985. In: Dixon. 1985: 37–66
Power and Cocking. 1971. Sci. Progr. (Oxford) 59: 181–198
Power et al. 1970. Nature 225: 1016–1018
Power et al. 1976. Nature 263: 500–502
Preil and Engelhardt. 1977. Acta Hort. 78: 203–208
Preil and Engelhardt. 1982. Gartenbauwiss. 47: 241–244
Preil et al. 1977. Z. Pflanzenzücht. 79: 167–171
Preil et al. 1983. Acta Hortic. 131: 345–351
Proft, de et al. 1985. Physiol. Plant. 65: 375–379

Quak. 1966. Landbouwk. Tijdschr. 78: 301–305
Quak. 1972. Proc. 18th Int. Hort. Congr. 3: 12–25
Quak. 1977. In Vitro 13: 194
Quoirin. 1974. Bull. Rech. Agron. Gembloux 9: 189–192

Rabéchault et al. 1968. Oléagineux 23: 233–238
Rabéchault et al. 1969. C.R. Acad. Sci. 268: 1728–1731
Rabéchault et al. 1972. Oléagineux 27: 249–254
Rabéchault et al. 1973. Oléagineux 28: 333–336
Radojevic and Kovoor. 1986. In: Bajaj. 1986: 65–86
Raghavan. 1966. Biol. Rev. 41: 1–58
Raghavan. 1976. Science 191: 388–389
Raghavan. 1976a. Experimental embryogenesis in vascular plants. Acad. Press, London.
      1976: 1–603
Raghavan. 1977. In: Reinert and Bajaj. 1977: 375–397
Raghavan. 1980. Int. Rev. Cytol. 11B: 209–240
Raghavan. 1986. Embryogenesis in angiosperms. Cambridge University Press, Cambrdige:
      1–303
Ramula et al. 1984. Plant Sci. Lett. 36: 79–86
Randall et al. 1983. Current topics in plant biochemistry and physiology. University Mis-
      souri, Columbia 2nd vol.: 1–350
Randolph. 1945. Bull. Am. Iris Soc. 97: 33–45
Randolph and Cox. 1943. Proc. Am. Soc. Hort. Sci. 43: 284–300
Rangan et al. 1968. HortScience 3: 226–227
Rangaswamy. 1977. In: Reinert and Bajaj. 1977: 412–425
Rangaswamy. 1981. In: Rao. 1981: 269–286
Rangaswamy and Shivanna. 1967. Nature 216: 927–929

Rangaswamy and Shivanna. 1971. J. Indian Bot. Soc. 50A: 286–296

Rao. 1965. Phyton 22: 165–167

Rao. 1977. In: Reinert and Bajaj. 1977: 44–69

Rao. 1981. Tissue culture of economically important plants. Costed and ANBS: 1–307

Rao and Rangaswamy. 1972. Bot. Gaz. 133: 350–355

Rao et al. 1980. Plant tissue culture, genetic manipulation and somatic hybridization of plant cells. Proc. Nat. Symp. Bhabba Atomic Research Centre, Bombay, India: 1–436

Rappaport. 1954. Bot. Rev. 20: 201–225

Raquin. 1986. Int. Congr. Plant Tissue Cell Culture 6: 123

Rashid. 1983. Physiol. Plant. 58: 544–548

Rathore and Goldsworthy. 1985. Bio/Technology 3: 1107–1109

Raveh et al. 1973. In Vitro 9: 216–222

Read et al. 1986. HortSci. 21: 796

Redéi and Redéi. 1955. Acta Bot. Acad. Sci. Hung. 2: 183–186

Redenbaugh et al. 1986. HortSci. 21: 819

Reeves. 1986. Span 29: 32–33

Refaat et al. 1984. Z. Pflanzenzücht. 93: 137–146

Reinert. 1958. Naturwiss. 45: 344–345

Reinert. 1959. Planta 53: 318–333

Reinert. 1973. In: Street. 1973: 338–355

Reinert and Bajaj. 1977. Plant cell, tissue, and organ culture. Springer Verlag, Berlin: 1–803

Reinert and Bajaj. 1977a. In: Reinert and Bajaj. 1977: 251–267

Reinert and Besemer. 1967. Wiss. Z. Univ. Rostock Math. Nat. Reihe 16: 599–604

Reinert and Mohr. 1967. Proc. Am. Soc. Hort. Sci. 91: 664–671

Reinert and Yeoman. 1982. Plant cell and tissue culture. A laboratory manual. Springer Verlag, Berlin: 1–83

Reinhard. 1974. In: Street. 1974: 433–459

Reinhard and Alfermann. 1980. Adv. Biochem. Engen. 16(1): 49–84

Reisch. 1983. In: Evans et al. 1983: 748–770

Reuther. 1978. Gartenbauwiss. 43: 1–10

Reuther. 1985. In: Schäfer-Menuhr. 1985: 1–14

Reuther and Sonneborn. 1981. Zierpflanzenbau 21: 478–479

Reyburn. 1978. Am. Orchid Soc. Bull. 47: 798–802

Rijven. 1952. Acta Bot. Neerl. 1: 157–200

Ringe and Nitsch. 1968. Plant Cell Physiol. 9: 639–652

Robb. 1957. J. Exp. Bot. 8: 348–352

Robbins. 1922. Bot. Gaz. 73: 376–390

Roberts et al. 1984. Plant Tissue Culture Letters 1: 22–24

Roest. 1977. Acta Hort. 78: 349–359

Roest. 1980. In: Doré. 1980: 186–191

Roest and Bokelmann. 1976. Vakbl. Bloem. 31(9): 36–37

Roest and Bokelmann. 1980. Potato Res. 23: 167–181

Romberger and Tabor. 1971. Am. J. Bot. 58: 131–140

Rosati et al. 1980. J. Am. Soc. Hort. Sci. 105: 126–129

Rosevear and Lambe. 1985. Adv. Biochem. Engineering Biotechn. 31: 37–58

Rossini and Nitsch. 1966. C.R. Acad. Sci. 263: 1379–1382

Roth and Lark. 1984. Theor. Appl. Genet. 68: 421–431

Rudelle. 1977. In: Gautheret. 1977: 232–237

Rugini and Wang. 1986. Int. Congr. Plant Tissue Cell Culture Abstr. 6: 374

Rugini et al. 1986. HortSci. 21: 804

Russell and McCown. 1986. Int. Congr. Plant Tissue Cell Culture Abstr. 6: 49

Rutkowski. 1983. Am. Orchid Soc. Bull. 52: 50

Rybczinski et al. 1983. Genet. Pol. 24: 1-8

Sachs. 1880-1882. Archiv. Bot. Inst. Würzburg 2: 453 and 689

Sacristan and Melchers. 1969. Molec. Gen. Genet. 105: 317-333

Sakai. 1984. Hortic. Rev. 6: 357-372

Sakai and Sugawara. 1973. Plant Cell Physiol. 14: 1201-1204

Sanders and Ziebur. 1963. In: Maheswari. 1963: 297-325

Sanders et al. 1975. Endomycorrhizas. Acad. Press, London: 1-626

San Noeum. 1976. Ann. Amél. Plantes 26: 751-754

Sastri. 1963. Morphogenesis in plant tissue cultures. Plant tissue and organ culture. A symposium (Delhi): 105-107

Saxena et al. 1986. Planta 168: 29-35

Schaeffer. 1974. Abstr. 3rd Int. Congr. Plant Tissue and Cell Culture, Leicester: 71

Schäfer-Menuhr. 1985. In vitro techniques. Propagation and long term storage. M. Nijhoff, Dordrecht, the Netherlands: 1-194

Schenck and Röbbelen. 1982. Z. Pflanzenzücht. 89: 278-288

Schenk and Hildebrandt. 1972. Can. J. Bot. 50: 199-204

Schieder. 1975. Z. Pflanzenphysiol. 76: 462-466

Schieder. 1976. Mol. Gen. Genet. 149: 251-254

Schieder. 1980. In: Rao et al. 1980: 216-227

Schilperoort. 1983. In: Linskens et al. 1983: 97-136

Schilperoort. 1984. In: Lange et al. 1984: 251-285

Schlechter. 1980. Die Orchideen Band II (3-5): 158-211

Schmidt and Keller. 1986. Int. Congr. Plant Tissue Cell Culture Abstr. 6: 122

Schneider and McKensie. 1986. Int. Congr. Plant Tissue Cell Culture Abstr. 6: 374

Schneider-Moldrickx. 1983. Acta Hort. 131: 163-170

Schraudolf and Reinert. 1959. Nature 184: 465-466

Scowcroft. 1985. In: Hohn and Dennis. 1985: 217-245

Seeliger. 1959. Flora 148: 218-254

Seemüller and Merkle. 1984. Gartenbauwiss. 49: 227-230

Seeni and Gnanam. 1981. Can. J. Bot. 59: 1941-1943

Seibert. 1976. Science 191: 1178-1179

Seibert et al. 1975. Plant Physiol. 56: 130-139

Semal. 1986. Somaclonal variations and crop improvement. M. Nijhoff, Dordrecht: 1-277

Sen and Giles. 1983. Plant cell culture in crop improvement. Plenum Press, New York: 1-502

Sharp et al. 1979. Plant cell and tissue culture. Principles and applications. Ohio State Univ. Press, Columbus: 1-892

Sharp et al. 1982. Proc. 5th Int. Congr. Plant Tissue Cell Culture 5: 759-762

Sharp et al. 1984. Handbook of plant cell culture. Vol. 2. Crop species. MacMillan Publ. Comp., New York: 1-644

Shepard et al. 1980. Science 208: 17-24

Shigematsu. 1974. Jap. Soc. Hort. Sci. Symposium: 302-303 and 348-349

Shigematsu and Matsubara. 1972. J. Jap. Soc. Hort. Sci. 41: 196-200

Shigenoby and Sakamoto. 1981. Jap. J. Genet. 56: 505-517

Shillito et al. 1983. Plant Cell Rep. 2: 244

Shillito et al. 1986. Newsletter I.A.P.T.C. 48: 5-15

Short and Roberts. 1986. Int. Congr. Plant Tissue Cell Culture Abstr. 6: 432

Siderov et al. 1981. Nature 294: 87-88

338

Silvoy et al. 1984. Soybean Genet. Newslett. 11: 80–82

Singh. 1981. Am. Orchid Soc. Bull. 50: 416–418

Singh and Prakash. 1984. Scientia Hort. 24: 385–390

Skene and Barlass. 1983. Z. Pflanzenzücht. 90: 130–135

Skene and Barlass. 1983a. J. Exp. Bot. 34: 1271–1280

Skirm. 1942. J. Hered. 33: 211–215

Skirvin et al. 1986. Plant Cell Rep. 5: 292–294

Skoog. 1944. Am. J. Bot. 31: 19–24

Skoog and Miller. 1957. Symp. Soc. Exp. Biol. no. 11. The biological action of growth substances: 118–131

Skoog and Tsui. 1948. Am. J. Bot. 35: 782–787

Sladky and Havel. 1976. Biol. Plant. 18: 469–472

Sluis and Walker. 1985. Newsletter I.A.P.T.C. 47: 2–12

Smaal. 1980. Vakbl. Bloem. 35(46): 46–47, 49

Smith. 1975. Plant Propagator 21: 14

Smith. 1986. In: Bajaj. 1986: 274–291

Sneep. 1982. Plantenveredeling een vak apart. Rede Landbouwhogeschool, Wageningen: 1–22

Sneep. 1983. In: Linskens et al. 1983: 76–96

Snir and Erez. 1980. HortSci. 15: 597–598

Soede. 1979. Vakbl. Bloem. 34(19): 30–31

Sree Ramulu et al. 1986. In: Anonymous. 1980: 77–82

Staba. 1977. In: Reinert and Bajaj. 1977: 694–702

Staba. 1980. Plant tissue culture as a source of biochemicals. CRC Press, Boca Raton, Florida: 1–285

Staba. 1985. J. Natural Products 48: 203–209

Staden, van and Drewis. 1975. Physiol. Plant. 34: 106–109

Staritsky and Hasselt, van. 1980. Proc. 9th Int. Scient. Colloq. Coffee 2: 597–602

Staritsky et al. 1983. Acta Bot. Neerl. 35: 491

Steffen et al. 1986. Theor. Appl. Gen. 72: 135–140

Stehsel and Caplin. 1969. Life Sci. 8: 1255–1259

Steward. 1958. Am. J. Bot. 45: 709–713

Steward et al. 1964. Science 143: 20–27

Stewart and Merwe, van der. 1982. Proc. 10th World Orchid Conf. Durban, South Africa: 263–277

Stichel. 1959. Planta 53: 293–317

Stimart and Ascher. 1981. Scientia Hortic. 14: 165–170

Stohs. 1980. Adv. Biochem. Engen. 16(1): 85–107

Stolz. 1967. Proc. Am. Soc. Hort. Sci. 64: 13–14

Stolz. 1971. J. Am. Soc. Hort. Sci. 96: 681–684

Stone. 1963. Ann. Appl. Biol. 52: 199–209

Stoutemeyer and Britt. 1965. Am. J. Bot. 52: 805–810

Stoutemeyer and Britt. 1970. BioSci. 20: 914

Straub. 1976. Vorträge Rhein.-Westf. Akad. Wiss. N257: 17–26

Straus. 1954. Am. J. Bot. 41: 833–839

Strauss and Reisinger. 1976. Am. Orchid Soc. Bull. 45: 722–723

Street. 1973. Plant tissue and cell culture. Bot. Monogr. Vol. II. Blackwell Scient. Publ., London: 1–503. Second edition in 1977: 1–540

Street. 1974. Tissue culture and plant science 1974. Acad. Press, New York: 1–502

Street. 1979. In: Sharp et al. 1979: 123–153

Stringham. 1977. Plant Sci. Lett. 9: 115–119
Strullu et al. 1984. Acad. Agric. France 70: 1331–1337
Styer. 1985. In: Henke et al. 1985: 117–130
Styer and Chin. 1983. Hortic. Rev. 5: 221–277
Sunderland. 1973. In: Street. 1973: 161–190
Sunderland. 1973a. In: Street. 1973: 205–239
Sunderland. 1978. In: Anonymous. 1978c: 65–86
Sunderland. 1980. In: Ingram and Helgeson. 1980: 33–40
Sunderland and Dunwell. 1974. In: Street. 1974: 141–167
Sutter. 1985. Ann. Bot. 55: 321–329
Sutter and Hutzell. 1984. Scientia Hortic. 23: 303–312
Syono. 1965a. Plant Cell Physiol. 6: 371–392
Syono. 1965b. Plant Cell Physiol. 6: 393–401
Syono. 1965c. Plant Cell Physiol. 6: 403–419

Tabata. 1977. In: Barz et al. 1977: 3–16
Takayama and Misawa. 1982. In: Fujiwara. 1982: 681–682
Takayama et al. 1986. Int. Congr. Plant Tissue Cell Culture Abstr. 6: 449
Takebe. 1976. Newsl. I.A.P.T.C. 19: 2–6
Takebe et al. 1971. Naturwiss. 58: 318–320
Tanaka and Sakanishi. 1977. Am. Orchid Soc. Bull. 46: 733–737
Tanaka and Sakanishi. 1985. Bull. Univ. Osaka Prefecture B 37: 1–4
Tangley. 1986. Bioscience 36: 414–420
Tateno et al. 1982. In: Fujiwara. 1982: 787–788
Teo. 1976–1978. Malayan Orchid Rev. 13: 56–59
Téoulé. 1983. C.R.Acad.Sci.Paris 297: 13–16
Theiler. 1971. Schweiz. Landwirtsch. Forschung 10: 65–93
Theiler. 1981. Obstbau 6: 97–103
Thévenin and Doré. 1976. Ann. Amél. Plantes 26: 655–674
Thomale. 1957. Die Orchideen. Ulmer, Stuttgart: 1–211
Thomas. 1984. Plant Mol. Biol. Rep. 2: 46–61
Thomas and Davey. 1975. From single cells to plants. Wykeman Publ., London: 1–172
Thomas and Street. 1970. Ann. Bot. 34: 657–669
Thomas et al. 1979. Z. Pflanzenzücht. 82: 1–30
Thorpe. 1978. Frontiers of plant tissue culture 1978. Proc. 4th Int. Congress Plant Tissue Cell Culture. Calgary, Canada: 1–556
Thorpe. 1981. Plant tissue culture. Methods and applications in agriculture. Acad. Press, New York: 1–379
Tisserat and Vandercook. 1985. Plant Cell Tissue Organ Culture 5: 107–117
Tisserat and Vandercook. 1986. Am. Orchid Soc. Bull. 55: 35–42
Tisserat et al. 1979. Hort. Rev. 1: 1–78
Toledo et al. 1980. C.R. Acad. Sci. D290: 539–542
Toponi. 1963. C.R. Acad. Sci. 257: 3030–3033
Tramier. 1965. C.R. Séanc. Acad. Agric. France 51: 918–922
Tremblay et al. 1986. In: Bajaj. 1986: 87–100
Tripathi. 1974. Rev. Cytol. Biol. Vég. 37: 1–106
Truong-Andre and Demarly. 1984. Z. Pflanzenzücht. 92: 309–320
Tukey. 1933a. J. Hered. 24: 7–12
Tukey. 1933b. Bot. Gaz. 94: 433–468
Tukey. 1935. Proc. Am. Soc. Hort. Sci. 32: 313–322
Tulecke. 1953. Science 117: 599–600

Turner. 1983. Australian Hortic. 81: 95–99

Vacin and Went. 1949. Bot. Gaz. 110: 605–613
Vajrabhaya. 1977. In: Arditti. 1977: 179–201
Vajrabhaya and Chaichareon. 1974. Abstr. 3rd Int. Congr. Plant Tissue and Cell Culture, Leicester: 280
Valmayor and Sagawa. 1967. Am. Orchid Soc. Bull. 36: 766–769
Vardi et al. 1975. Plant Sci. Lett. 4: 231–236
Vasil. 1978. Newsl. I.A.P.T.C. 26: 2–10
Vasil. 1980. Int. Rev. Cytol. 11a: 1–253
Vasil. 1980a. Int. Rev. Cytol. 11b: 1–257
Vasil. 1984. Cell culture and somatic cell genetics. Volume 1. Laboratory procedures and their applications. Orlando, Florida: 1–825
Vasil. 1986. In: Semal. 1986: 108–116
Vasil and Hildebrandt. 1965. Science 150: 889–892
Venketeswaran and Partanen. 1966. Radiation Bot. 6: 13–20
Verhagen et al. 1986. Int. Congr. Plant Tissue Cell Culture Abstr. 6: 58
Verna and Collins. 1984. Theor. Appl. Genet. 69: 187–192
Vienken et al. 1981. Physiol. Plant. 53: 64–70
Vij et al. 1984. Bot. Gaz. 145: 210–214
Vilaplana-Marshall and Mullins. 1986. HortSci. 21: 734

Waes, van and Debergh. 1986. Physiol. Plant. 66: 435–442
Walkey. 1978. In: Thorpe. 1978: 245–254
Walkey. 1980. In: Ingram and Helgeson. 1980. 109–117
Walkey and Cooper. 1976. Ann. Appl. Biol. 84: 425–428
Walkey et al. 1974. J. Hort. Sci. 49: 273–275
Ward and Vance. 1968. J. Exp. Bot. 19: 119–124
Wardle et al. 1983a. J. Am. Soc. Hort. Sci. 108: 386–389
Wenzel. 1980. In: Rao et al. 1980: 68–78
Wenzel. 1983. Acta Hort. 131: 15–22
Wenzel and Foroughi-Wehr. 1984. In: Vasil. 1984: 293–301
Wersuhn and Fritze. 1985. Biochemie Physiol. Pflanzen 180: 79–83
Weston and Street. 1968. Ann. Bot. 32: 521–529
White. 1934. Plant Physiol. 9: 585–600
White. 1934a. Phytopath. 24: 1003–1011
White. 1939. Am. J. Bot. 26: 59–64
Wickson en Thimann. 1958. Physiol. Plant. 11: 62–74
Widholm. 1978. Newsl. I.A.P.T.C. 23: 2–6
Widholm. 1982. Proc. 5th Int. Congr. Plant Tissue Cell Culture 5: 609–612
Williams. 1978. New Zeal. J. Bot. 16: 499–506
Williams and Lautour, de. 1980. Bot. Gaz. 141: 252–257
Williams and Maheswaran. 1986. Ann. Bot. 57: 443–462
Williams et al. 1985. HortSci. 20: 1052–1053
Wimber. 1963. Am. Orchid Soc. Bull. 32: 105–107
Winton. 1968. Science 160: 1234–1235
Withers. 1980. Tissue culture storage for genetic conservation. Rep. Int. Board Plant Gen. Res. FAO, Rome: 1–91
Withers. 1980a. In: Vasil. 1980a: 101–136
Withers. 1982. Institutes working on tissue culture for genetic conservation. Rep. Int. Board Plant Gen. Res. FAO, Rome: 1–104

Withers. 1983. Span 26: 72–74
Withers. 1986. In: Withers and Alderson. 1986: 261–276
Withers. 1986a. In: Yeoman. 1986: 96–140
Withers and Alderson. 1986. Plant tissue culture and its agricultural applications. Butterworth, London: 1–526
Withers and Street. 1977. In: Barz et al. 1977: 226–244
Withner. 1974. The Orchids. J. Wiley, New York: 1–604
Woltering. 1986. Acta Bot. Neerl. 35: 50
Wood. 1985. In: Dixon. 1985: 193–214
Wullems et al. 1982. In: Fujiwara. 1982: 505–506
Wullems et al. 1986. In: Evans et al. 1986: 197–220
Wurm. 1960. Flora 149: 43–76

Yamaguchi and Nakajima. 1974. In: Plant Growth Substances 1973. Hirokawa Publ. Co., Tokyo, Japan: 1121–1127
Yeoman. 1986. Plant cell culture technology. Blackwell Scient. Publ., Oxford: 1–375
Yeung et al. 1981. In: Thorpe. 1981: 253–271
Yoshida and Ogawa. 1983. Techn. Bull. Food Fertilizer Techn. Center, Taiwan 73: 1–20
Yoshida et al. 1973. Plant Cell Physiol. 14: 329–339
Young et al. 1981. Plant Sci. Lett. 34: 203–210

Zaenen et al. 1974. J. Molec. Biol. 86: 109–127
Zagaja. 1962. Hort. Res. 2: 19–34
Zenk. 1976. Vorträge Rhein. Westf. Akad. Wiss. N257: 27–55
Zenkteler. 1971. Acta Soc. Bot. Poloniae 40: 305–313
Zenkteler. 1980. In: Vasil. 1980a: 137–156
Zenkteler and Straub. 1979. Z. Pflanzenzücht. 82: 36–44
Ziebur and Brink. 1951. Am. J. Bot. 38: 253–256
Zimmer. 1980. In: Schlechter. 1980: 158–211
Zimmer et al. 1981. Deut. Gartenbau 36: 1195–1197
Zimmerman. 1985. In: Henke et al. 1985: 165–177
Zimmerman et al. 1986. Tissue culture as a plant production system for horticultural crops. M. Nijhoff, Dordrecht: 1–371
Zimmermann. 1982. Biochim. Biophys. Acta 694: 227–277
Ziv. 1986. In: Withers and Alderson. 1986: 187–196
Ziv and Halevy. 1983. Hort. Sci. 18: 434-436

* After the manuscript was taken into production the following series of three volumes was published,
Bonga and Durzan. 1987. Cell and Tissue Culture in Forestry.
  Vol. 1: General Principles and Biotechnology: 1–422
  Vol. 2: Specific Principles and Methods: Growth and Developments: 1–447
  Vol. 3: Case Histories: Gymnosperms, Angiosperms and Palms: 1–416
  Martinus Nijhoff Publishers, Dordrecht, the Netherlands.

Acknowledgements

The Author and Publishers are grateful that permission was granted on the following illustrations:

Fig. 2.5 :	Vakblad Bloemisterij	Misset, Doetinchem
Fig. 3.1 :	Landbouwkundig Tijdschrift	Koninklijk Genootschap voor Landbouwwetenschap
Fig. 4.1 :	Vakblad Bloemisterij	Misset, Doetinchem
Fig. 6.9 :	Vakblad Bloemisterij	Misset, Doetinchem
Fig. 6.11:	Zeitschrift Pflanzenphysiologie	G. Fischer Verlag, Stuttgart
Fig. 6.14:	Landbouwkundig Tijdschrift	Koninklijk Genootschap voor Landbouwwetenschap
Fig. 12.3 :	Netherlands Journal Agricultural Science	Koninklijk Genootschap voor Landbouwwetenschap
Fig. 13.3 :	Netherlands Journal Agricultural Science	Koninklijk Genootschap voor Landbouwwetenschap
Fig. 13.6 :	Zeitschrift Pflanzenphysiologie	G. Fischer Verlag, Stuttgart
Fig. 16.3 :	Proc. Am. Soc. Hort. Sci.	American Society of Horticultural Science
	L. Nickell	Arkansas, U.S.A.
Fig. 18.1 :	Révue Géneral Botanique	Editions Belin, Paris
Fig. 18.2 :	Révue Géneral Botanique	Editions Belin, Paris
Fig. 18.3 :	Landbouwkundig Tijdschrift	Koninklijk Genootschap voor Landbouwwetenschap
Fig. 18.4 :	American Orchid Society Bulletin	American Orchid Society
Fig. 19.1 :	Landbouwkundig Tijdschrift	Koninklijk Genootschap voor Landbouwwetenschap
Fig. 19.2 :	Landbouwkundig Tijdschrift	Koninklijk Genootschap voor Landbouwwetenschap
Fig. 19.5 :	Proc. 18th Intern. Hort. Congr.	International Society for Horticultural Science
Fig. 20.1 :	Acta Horticulturae	International Society for Horticultural Science

Fig. 20.3 :	Acta Horticulturae	International Society for Horticultural Science
Fig. 24.3 :	Euphytica	Hoofdredacteur Euphytica, Wageningen
Fig. 24.8 :	Canadian J. Genet. Cytology	The Genetic Society of Canada
		O.L. Gamborg
		Fort Collins, U.S.A.
Fig. 24.9 :	Vakblad Bloemisterij	Misset, Doetinchem
Fig. 25.2 :	Vakblad Bloemisterij	Misset, Doetinchem
Fig. 25.4 :	Planta Medica	Hippocrates Verlag, Stuttgart
Fig. 25.7 :	Groente en Fruit	Hoofdredacteur Groente en Fruit, Den Haag

First Edition of the Dutch edition of this book (1975)   Thieme, Zutphen
Second Edition of the Dutch edition of this book (1985)   R.L.M. Pierik, Wageningen

Plant Cell, Tissue and Organ Culture

An International Journal on the Cell Biology of Higher Plants

Managing Editorial Board

Donald K. Dougall (Chairman), *University of Tennessee, Knoxville, USA;* **Michael W. Fowler**, *University of Sheffield, UK*

Plant Cell, Tissue and Organ Culture is an international monthly journal that publishes original results of studies on the various aspects of the cell biology of higher plants – including fundamental, methodological, and technical aspects. Topics include growth and development; technical aspects; anatomical, histological, and ultrastructural aspects; physiology and biochemistry; genetics and breeding; phytopathological aspects; and biotechnology.

Subscription Information   **ISSN** 0167–6857
1987, Volume 8–11 (12 issues)
Institutional rate: Dfl. 672.00/US$280.00 incl. postage/handling
Private rate: Dfl. 260.00/US$150.00 incl. postage/handling
Special rate for TCA members: US$74.00 incl. postage/handling

Private subscriptions should be sent direct to the publishers

Back Volume(s) Available   **Price per Volume**
Volumes 1–7 (1981–1986)   excl. postage
Dfl. 155.00/US$62.00
(Vols. 1–3)
Dfl. 125.00/US$47.00
(Vols. 4–5)
Dfl. 134.00/US$47.00
(Vols. 6–7)

Abstracts   **ISSN** 0920–6442

1987, Volume 2 (3 issues)
Institutional rate: Dfl. 176.00/US$73.00 incl. postage/handling
Private rate: Dfl. 80.00/US$32.50 incl. postage/handling

Martinus Nijhoff Publishers

Kluwer academic publishers group

P.O. Box 989   101 Philip Drive   Falcon House
3300 AZ Dordrecht   Norwell,   Queen Square
The Netherlands   MA 02061, U.S.A.   Lancaster, LA1 1RN, U.K.

Tissue Culture as a Plant Production System for Horticultural Crops

Conference on Tissue Culture as a Plant Production System for Horticultural Crops, Beltsville, MD, October 20–23, 1985

edited by
RICHARD H. ZIMMERMAN
ROBERT J. GRIESBACH
FREDDI A. HAMMERSCHLAG
ROGER H. LAWSON

CURRENT PLANT SCIENCE AND BIOTECHNOLOGY IN AGRICULTURE 2

1986, 381 pp.   ISBN 90–247–3378–2
Hardbound Dfl. 105.00/£32.75/US$ 45.00

Martinus Nijhoff Publishers

Micropropagation of horticultural crops has played an increasingly significant role in production of horticultural crops in recent years. Recent trends in construction and enlargement of tissue culture laboratories in the US show that numbers of plants produced could increase by two to three times within the next two years with some laboratories capable of a much greater increase. Since the magnitude of the changes in the industry are so great and the range of crops being produced is increasing so rapidly, a conference was organized to permit researchers and commercial users to exchange information on *in vitro* propagation of horticultural crops. Topics covered at the conference included basic principles of plant tissue culture and applications to plant improvement and propagation, new technology, genetic stability of tissue cultured plants, pathogen detection and elimination, plant quarantine and international shipment of tissue cultured plants, economics of tissue culture propagation, laboratory design, large scale production and propagation of specific groups of plants.

 Kluwer academic publishers group

P.O. Box 989   101 Philip Drive   Falcon House
3300 AZ Dordrecht   Norwell,   Queen Square
The Netherlands   MA 02061, U.S.A.   Lancaster, LA1 1RN, U.K.